# *Rocking the Boat: Conserving Fisheries and Protecting Jobs*

ANNE PLATT McGINN

Sophie Chou, *Research Intern*

Jane A. Peterson, *Editor*



WORLDWATCH PAPER 142
June 1998

**THE WORLDWATCH INSTITUTE** is an independent, nonprofit environmental research organization in Washington, DC. Its mission is to foster a sustainable society in which human needs are met in ways that do not threaten the health of the natural environment or future generations. To this end, the Institute conducts interdisciplinary research on emerging global issues, the results of which are published and disseminated to decisionmakers and the media.

**FINANCIAL SUPPORT** for this research is provided by the Curtis and Edith Munson Foundation. Financial support for the Institute is provided by the Geraldine R. Dodge Foundation, the Ford Foundation, the Foundation for Ecology and Development, the William and Flora Hewlett Foundation, W. Alton Jones Foundation, John D. and Catherine T. MacArthur Foundation, Charles Stewart Mott Foundation, the Curtis and Edith Munson Foundation, David and Lucille Packard Foundation, Rasmussen Foundation, Rockefeller Brothers Fund, Rockefeller Financial Services, Summit Foundation, Surdna Foundation, Turner Foundation, U.N. Population Fund, Wallace Genetic Foundation, Wallace Global Fund, Weeden Foundation, and the Winslow Foundation. The Institute also receives financial support from the Friends of Worldwatch and from our Council of Sponsor members: Tom and Cathy Crain, Toshishige Kurosawa, Kazuhiko Nishi, Roger and Vicki Sant, Robert Wallace, and Eckart Wintzen.

**THE WORLDWATCH PAPERS** provide in-depth, quantitative and qualitative analysis of the major issues affecting prospects for a sustainable society. The Papers are written by members of the Worldwatch Institute research staff and reviewed by experts in the field. Published in five languages, they have been used as concise and authoritative references by governments, nongovernmental organizations, and educational institutions worldwide. For a partial list of available Papers, see back pages.



Library of Congress Catalog Number 98-060965

ISBN 1-878071-44-0

Printed on 100-percent non-chlorine bleached, partially recycled paper.

# *Table of Contents*

**ACKNOWLEDGMENTS:** I thank Peter Auster, Menakam Ben-Yami, Peter Edwards, Veravat Hongskul, Pamela Mace, Dan MacMeekin, Matteo Milazzo, Bill Mott, Norman Myers, Chris Newton, Elliott Norse, Cynthia Suchman, Alan White, and Don Woodworth for their invaluable comments and contributions to the paper. Special thanks to Maurizio Perotti of FAO for his help with numerous data requests; to Don Hinrichsen for his ideas and input which were critical to this paper; and to the staff at ICLARM, who hosted my visit to the Philippines in January and provided me with a wealth of information. I also acknowledge the contributions of many people in the field who have influenced my work and helped forward the cause of marine conservation.

Thanks to all my colleagues at Worldwatch for key pieces of information, suggestions, and encouragement. Special thanks to Peter Weber for getting me hooked on fisheries in the first place; to Hilary French, for helpful reviews and overall guidance; Janet Abramovitz and Brian Halweil for thoughtful feedback on the first draft; Sophie Chou, for help with tables, figures, and overall content; Ashley Mattoon for checking the endnotes; Amy Warehime, for guidance from cradle to grave; Mary Caron and Dick Bell, for help with the outreach; Liz Doherty for adjusting her schedule for the sake of the best possible product, and for her wit and sensitivity; and Jane Peterson, for editing the style and substance of the paper with a human touch. I give heartfelt thanks to my husband, Joe, for everything.

**ANNE PLATT McGINN** is a research associate at the Worldwatch Institute where she researches fisheries, oceans, and environmental health issues. She is a regular contributor to the Institute's annual publications, *State of the World,* and *Vital Signs*. She wrote Worldwatch Paper 129, *Infecting Ourselves: How Environmental and Social Disruptions Trigger Disease,* and has authored articles in *World Watch* magazine on topics including, aquaculture, duckweed, golf, water and sanitation, tobacco, and tuberculosis. Prior to joining the Institute in April 1993, she worked as a research assistant at the Environmental Law Institute. Ms. McGinn graduated from Williams College in 1991 with a B.A. in political science and a concentration in environmental studies.

## *Introduction*

In the port of Prince Rupert, British Columbia, late in July 1997, angry commercial fishers blockaded a U.S. ferry bound for Alaska to protest excessive fishing by their American counterparts. For three days, some 300 innocent bystanders onboard the ferry witnessed firsthand the increasingly common symptoms of global overfishing: mounting social disruption, economic pressures, and, on occasion, outright violence.[1]

The crisis began to erupt in spring 1997, when U.S. and Canadian representatives of the Pacific Salmon Commission broke off their annual negotiations over the amount of catch to be permitted. Taking advantage of the absence of limits, Alaskan fishers proceeded to take more than three times as much sockeye salmon as had ever been allowed under the Pacific Salmon treaty. Desperate to hold onto their share of the beleaguered salmon—which migrate from Alaskan waters nearly 600 kilometers south toward Vancouver Island and Seattle during the summer—the Canadian fishers responded with public protests and threats of violence.[2]

The roots of the crisis run deep. One primary cause is that salmon are disappearing, due in large part to the destruction of salmon runs and the loss of upstream habitat areas to clear-cutting, dam building, and urbanization. In the northwestern United States, Pacific salmon have vanished from about 40 percent of their breeding range, and

We gratefully acknowledge the financial support of this research by the Curtis and Edith Munson Foundation.

more than 300 distinct salmon populations are at risk of extinction.[3]

A second key problem stems from the fact that a growing share of salmon now come from netcages in coastal waters located at the mouths of rivers. Local fish farmers have reportedly "dumped" cheaper farmed salmon on the market just before the wild species become available for capture, thereby distorting prices and further frustrating fishers' attempts to make a living. Imports of farmed salmon from Chile and Norway also compete with dwindling supplies of salmon caught in the Northwest, creating such a dependable supply that there is little chance that prices will rise again. Faced with declining wild catches and falling prices, salmon fishers are being squeezed out. If they go, so too will an important part of Pacific Northwest culture—communities founded on salmon fishing and processing. One local inhabitant called salmon "our statue of liberty."[4]

Seeking compensation for lost tourism dollars, Alaska has since sued the British Columbian fishers, while the government of British Columbia has sued the U.S. government for failing to curb excessive takings by Alaskan fishers. Efforts are being made to resolve the dispute peacefully, but with an agreement yet to emerge, future fishing seasons may bring more headlines of the 'salmon war' between these two normally friendly allies.[5]

Far from being an isolated incident, the recent flare-up in British Columbia typifies rising tensions worldwide. Fishers across the globe are feeling the effects of a bloated, inefficient industry and severe resource depletion. Last year alone, the United Nations reported more than 100 fishery disputes between individual nations. With fishing jobs at a premium and coastal communities bearing the brunt of social and economic stress, the world's fishers and fisheries are in crisis.[6]

Globally, fisheries—and the economic and social benefits they offer society—are under siege. By recent U.N. estimates, a majority of marine fish stocks and all of the world's primary fishing grounds have reached peak production and are in decline: landings of the most commercially valuable species have dropped by one fourth since 1970. Overall catches have only grown at about 1 percent annually during the 1990s, compared to 3 percent in the 1980s. Fish of lesser quality and value now fill fishers' nets. In the course of depleting prized species, fishers are quickly unraveling the food chain and disrupting marine ecosystems.[7]

Wielding more than twice the gear and equipment needed to extract available resources, fishers have wiped out many individual fisheries and prompted a freefall in global fish stocks—and their own profits. The sector barely stays afloat, even as tens of billions of dollars in global subsidies are poured into the industry, encouraging huge factory ships to haul in more fish and impoverished farmers to become fishers. About one third of all revenues from fisheries come not from wild catches, but from government coffers in the form of subsidized loans, preferential tax rates, and other means of economic support.[8]

At the same time that fish populations are being overexploited, the areas that serve as natural fish farms for wild stocks—coral reefs, tidal estuaries, and oceanfloor environments—are being ravaged by indiscriminate fishing gear and destructive fishing methods. Pollution from cities, farms, and industries also takes a toll on the underwater habitats that fish use to replenish their stocks.[9]

With nearly half of all fish caught today traded internationally, distant markets and foreign economic pressures also play a role in the depletion of global fisheries. Because of overfishing in temperate, northern waters, many previously self-sufficient nations, including the United States, now rely on imports from developing nations, where 85 percent of internationally traded fishery products originate. As fishers increasingly head to waters in the southern hemisphere, small-scale fishers must now compete with large vessels from North and South. Yet the smaller-scale fishers provide a majority of fish for the more than 1 billion people, predominantly in Asia and coastal developing countries, who rely on fish as their primary source of animal protein.

With a growing share of the catch exported, local prices of wild species are rising, so people in protein-deficient, low-income countries are hard pressed to meet their daily protein needs.[10]

As human demand for fish rises and marine catches stagnate, fishers and suppliers are increasingly looking to aquaculture, or fish farming, to fill the gap. In 1995, an estimated 20 percent of fish consumed worldwide were raised on a farm, compared to just 8 percent in 1984. Practiced in freshwater areas of China and South Asia for centuries, aquaculture offers significant resource benefits over marine fishing if, for example, the waste of one species is used as food for another, thereby closing the "organic loop," and if the farmed fish eat plants rather than other fish. Some fish farmers, however, have turned their back on these benefits by adding excessive amounts of feed, fertilizer, and even antibiotics to try to enhance output. In particular, the recent explosion in shrimp and salmon farming is putting more pressure on natural fisheries because vast quantities of ocean fish are processed into high-protein pellets to feed these carnivorous fish. And in the aquatic equivalent of "slash-and-burn" agriculture, some fish farms have supplanted wetlands, mangroves, and other habitat areas that are vital to the survival of wild species. Unless it is practiced in a environmentally sound manner, aquaculture can result in a net loss of fish—and protein. If it is done right, however, aquaculture can meet people's nutritional needs without drawing down natural fish stocks or their habitats.[11]

Based on the tenet of freedom of the seas dating back to the early 17th century, most fisheries have traditionally been open to all comers. But as growth in the fish catch has slowed and overcapacity in the world's fishing fleets has soared, open access—unlimited entry and use—has become untenable. The difficult question today is how to limit access. In the past couple of decades, governments have implemented a wide range of quotas and fishing moratoriums, but they have rarely consulted with the fishers themselves or allowed them to participate. Modern fishing agreements and the Convention on the Law of the Sea, which were established in part to make the industry more responsive to biological conditions, have failed to reconcile the conflicts between open access and exclusive rights over fishing. And in the face of declining resources, these agreements have actually aggravated competing claims for catches, as evidenced in British Columbia.[12]

In the process of establishing modern fisheries agreements and introducing some degree of limited access, leaders have largely ignored or replaced preexisting systems of community-based management that have prevailed on a limited scale in some areas. For example, sea tenure systems have sustained island peoples in the Pacific for thousands of years, while community-defined systems of limited access and entry have evolved over several generations in some coastal areas, where boundaries are easily defined and access is limited by family, type of gear, or season. Community-based systems often encompass a broader belief system that values the sea for its role in social history and cultural identity, as well as for food and sustenance, effectively deterring overfishing, and offering valuable lessons to policymakers today.[13]

Community-based systems of fisheries governance that persist in some coastal fisheries and tropical reef systems offer two key insights to sustainable fisheries management. First, the means of dividing up the waters should be based on equity, fairness, and need as determined by dependence on the resource and the best available scientific knowledge, not simply on economic might and political pressure; and second, successful management requires that a high degree of decisionmaking and accountability be shared by fishers and government officials.[14]

With tensions over fisheries increasing, demand for fish rising, and local food security at risk, the stakes are higher today than ever before. Accelerating economic and social desperation continues to drive fishers down the path of self-destruction, threatening to decimate world fishery resources. As a result, millions of fishers and hundreds of coastal communities worldwide are suffering from economic losses, political tensions, and social instability. Overcoming this legacy of mismanagement will not be easy.

A number of steps will help reform fisheries management and end the harmful practices that increasingly characterize humanity's relationship with fisheries. The most important elements are reducing fishing capacity and protecting habitat areas. The extent to which fisheries can recover depends on how quickly and to what degree redundant fishing fleets are retired; how soon fishers stop pulling in fish that are too young, too small, or too scarce; and whether they reduce waste, limit environmental damage inflicted in the process of fishing, and make better use of what is caught.

If the industry is to survive, fisheries management will need to make a fundamental shift away from managing fish supplies to managing the fishers and how they fish. The key is not deciding who takes how many fish, but how to balance the need to supply food and jobs with the needs of the resource. Until the requirements of fish and fishers are taken into consideration, there may be little to prevent blockades from turning into all-out gun battles. And unless fishers stop degrading the resource base and harming breeding populations, they will soon reach the point where it becomes too difficult or expensive to fish. Eventually they could cause biological extinction of important species.

Fortunately, efforts to promote needed changes are already under way in a number of places. From the grassroots level to the highest reaches of government, people have successfully negotiated new courses of action, utilizing many of the elements of successful sustainable fisheries management. Fishers and fishery officials alone cannot make the necessary hard choices; politicians and the public need to act together to reshape incentives and behavior in favor of fishing that is economically viable, socially diverse, and ecologically sustainable.

# *Fishing Down and Out*

In 1871, the U.S. government created its first federal conservation agency, the Commission of Fish and Fisheries, in response to the decline of fisheries off the coast of New England and in inland lakes. The Commission was charged with determining what caused the declines in this valuable food supply and whether or not "protective prohibitory or precautionary measures" should be adopted. The first Commissioner of Fisheries, internationally acclaimed scientist Spencer Fullerton Baird, emphasized that one could not understand fisheries without examining their food, habitat, predators, ocean conditions, and other factors. It is a message that still falls mainly on deaf ears.[15]

Fishery declines were nothing new even in 1871—in the mid-1800s, for example, drastic declines of whales had captured people's attention. But today, when the most valuable groundfish species (including cod, flounder, and hake in New England), are in far worse shape than during Baird's time, many people remain unaware of the threat. While human numbers climb and appetites for fish increase steadily, worldwide marine catches remain flat. And despite the greater magnitude of fishery problems today than during Baird's time, relatively little action is being taken to solve them. Scientists' calls for precaution and protective measures are largely ignored by policymakers, who focus on enhancing commerce, trade, and market supply, and look to extract as much from the sea as possible, with little regard for the effects of this exploitation on fish stocks or habitats.[16]

As a result of excessive exploitation and other abuses, most of the highly prized marine fisheries around the world are on the verge of collapse. The warning signs are clear: 11 of the world's 15 most important fishing areas and 60 percent of the major fish species are in decline, according to the U.N. Food and Agriculture Organization (FAO). Whereas no fish stocks were in urgent need of management in 1950, today a majority of the world's fisheries qualify for that dubious distinction, and require immediate action to reduce capacity and rehabilitate damaged resources.[17]

On the surface, it might appear to consumers shopping in well-stocked markets that there is no problem with the global fish supply. Between 1950 and 1996, world fish production—wild catches and farmed fish combined—increased dramatically, from 21 to 120 million tons. (See Figure 1.) In 1994, world fish production overtook production of beef and poultry combined. But these figures are deceptive, for world fish production has continued its steady climb upward largely due to the recent boom in aquaculture. Wild catches from marine and inland waters have expanded from 20 to 93 million tons since 1950—reaching a plateau of about 90 million tons during the last decade. Meanwhile, farmed fish production more than tripled between 1984 (the first year global aquaculture statistics were compiled by FAO) and 1996, from nearly 7 million tons to 26 million tons, and now accounts for one fifth of fish supplies worldwide. By the year 2000, this share is expected to rise to one fourth.[18]

Although world fish output reached an all-time high in 1997 and is expected to keep climbing because of continued growth in aquaculture, the era of rapidly expanding marine catches ended over two decades ago. From 1950 to 1970, marine catches grew at an annual rate of 6 percent—almost double the rate of population growth. But in the early 1970s, one of the world's largest fisheries by volume, the Peruvian anchovy, took a plunge that signaled the end of uninhibited growth in world landings. Jumping from near zero in the 1950s to 13 million tons in 1970, the catch of this one species comprised one fifth of the world's total at the time. But changes in the temperature of the Pacific Ocean due to El Niño, combined with overfishing, caused catches to collapse to less than 2 million tons in 1974. During the remainder of the 1970s and 1980s, expansion in marine catches slowed to 2.3 percent, dropping to near zero during the 1990s. Although the Peruvian anchovy recovered to 12 million tons in 1995, most other marine fish stocks have suffered a different fate.[19]

**FIGURE 1**

## Total World Fish Production, 1950–96

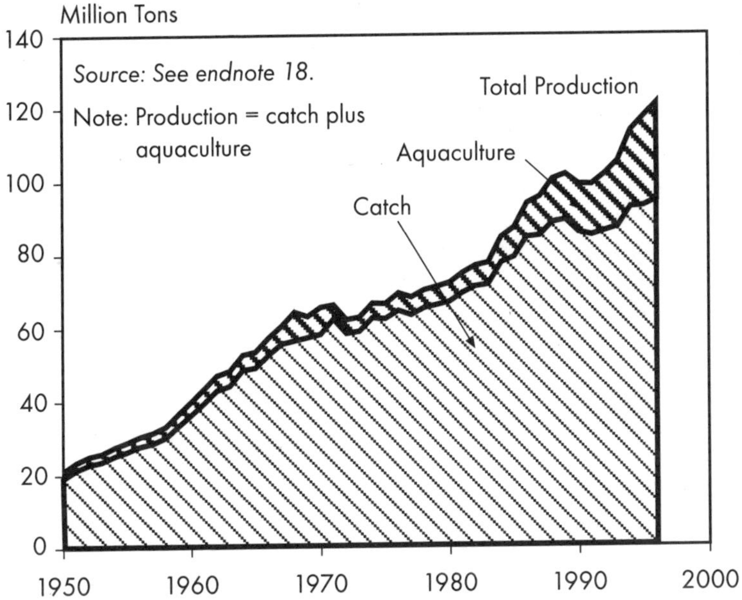


Many fish are on their last gasp. Shark populations off the southeastern U.S. coast have been slashed to less than half of their size in just the past 20 years. Japan's fishery production fell for the seventh year in a row—from 12 million tons in 1988 to less than 7 million tons in 1995—due to large declines in the sardine and mackerel catch. A look at particular fishing grounds reveals a similar picture. Fishers in the Northwest Atlantic have seen their bounty fall by 40 percent since the early 1970s, while fishers working in the Southeast Atlantic off the coast of Namibia and South Africa have experienced more than a 50 percent decline since that time.[20]

The situation is particularly troubling in the heavily fished waters of the Baltic and North seas. The International Council for the Exploration of the Seas, a Denmark-based scientific organization founded in 1911, has warned for years of overfishing in the region and has tried to persuade government officials to put fishery protection plans in place. But the council's efforts have largely failed. Since the 1960s, the North Sea mackerel fishery has declined by 80 percent, while the herring fishery, even though it was shut down entirely between 1977 and 1982, has never fully recovered. In March 1997, scientists and environmental ministers from 10 European Union (EU) nations and Norway issued an urgent warning—the latest in a series of such notices—that the herring-dependent cod fisheries in the North Sea were in danger of collapsing.[21]

Downward trends are also common in Asia. In the Gulf of Thailand, for example, commercial trawlers were introduced in the early 1960s. By 1982, fishers experienced an 80 percent drop in their daily catches. More recently, the average catch of fishers who harvest sea urchins (which are prized for their eggs) in Bolinao, Philippines, plummeted by more than 70 percent between 1989 and 1992, despite the fact that fisheries were closed down for three months per year to allow the species to reproduce.[22]

Fish are no longer safe from overexploitation, even in the most remote reaches of the sea where nutrients and species are widely dispersed. The fate of the orange roughy is a case in point. Pitted against modern vessels, high-tech gear, and seemingly unlimited capital, orange roughy are no match for sophisticated tracking systems and fishing nets that reach more than a kilometer below the surface of the South Pacific waters. Catches of this species plummeted by 70 percent in just six years. Because they need at least 25 years to mature to their breeding age, these long-lived deep sea dwellers could not replenish fast enough to compensate for losses of mature individuals, and their stocks dropped off. Perhaps more troubling than the rapid annihilation that can occur are the resulting reductions in the genetic diversity of the spawning populations, which make it more difficult for the species to adapt to future environmental changes. Species such as the orange roughy have been fished down to the point where future recoveries may be impossible.[23]

Fishers are now so efficient that they can, and do, wipe

out entire populations of fish, and then move on either to a different species or to a fishing area in some other part of the world. (See Figure 2.) Following the decline of groundfish stocks in the late 1980s and early 1990s, for instance, fishers still working in the Grand Banks region off the Atlantic coast of Canada started catching dogfish (a type of shark), skate, monkfish, and other species once considered trash. Similarly, U.S. fishers who targeted bluefin tuna off the mid-Atlantic coast in the 1960s switched to swordfish a decade later, and by the mid-1980s they had shifted their efforts to yellowfin tuna as stocks of the more desirable bluefin and swords dwindled.[24]

Because larger species feed on species lower down on the food chain, overfishing top predators such as cod or tuna

**FIGURE 2**

**Trends in Selected Open Access U.S. Fisheries, 1835–1995**

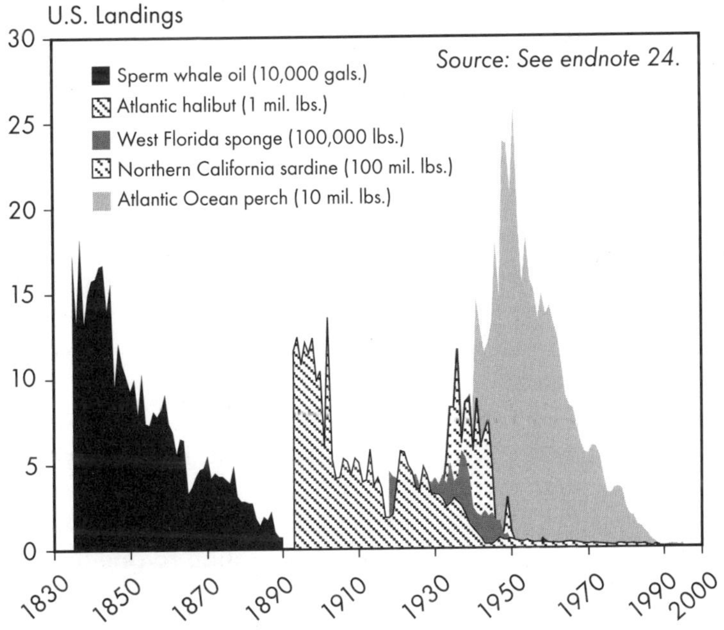

or shark triggers a slow-acting domino effect. At higher levels of the food chain, fish are bigger, but there are fewer of them than at lower levels, where species are smaller and more plentiful. Initially, the transition from high-level species to ones lower on the food chain brings new bounty. But unless the volume of fishing is reduced, the cycle of overfishing soon repeats itself with new prey: excessive fishing can trigger abrupt declines in these lower-level species, leaving fishers only steps away from the base of the food chain.[25]

As a result of overfishing of many species at the higher levels of the food chain, the composition of global catches has shifted to smaller, bonier, oily fish that eat lower down. The volume of *pelagic* fish (generally small, short-lived species that travel in schools in the open oceans) has increased from an estimated 6 million tons in 1950 to 44 million tons in 1994—from one third to one half of a growing world fish catch. During the 1980s, five low-value pelagic species—the Peruvian anchovy, South American pilchard, Japanese pilchard, Chilean jack mackerel, and Alaskan pollock (a semi-pelagic fish)—accounted for 73 percent of the increase in total capture. In sharp contrast, catches of four valuable *demersal* (bottom-dwelling) species—silver hake, haddock, Cape hake, and Atlantic cod—decreased by 67 percent between 1970 and 1992.[26]

Consequently, fishers now haul in species at a younger age and smaller size—either the juvenile stage of the same fish, or altogether different ones. Targeting young fish undermines future breeding populations and guarantees a smaller biological return in the coming years. For example, the average size of swordfish caught on longlines—fishing lines up to 50 kilometers long, equipped with thousands of baited hooks—has shrunk from 120 kilograms to 30 kilograms during the past 20 years. As a result, breeding populations of swordfish have declined by half, and the catch now consists of primarily small, immature fish. Consumers are essentially "eating the babies," according to one marine biologist.[27]

Declines in one species can also leave ecosystems vulnerable to invasive species by altering predator-prey relations. On the coral reefs of the Caribbean and the northern reaches of the Red Sea, for example, the overharvesting of triggerfish and pufferfish for souvenirs has sapped the health of the entire reef. As these fish declined, populations of their prey—sea urchins—exploded, damaging the coral by grazing on the protective layers of algae, and the health of coral reefs suffered, while the diving industry for tourists declined. Similarly, fishing removed top predators in Kenya's coral reefs, leading to an increase in the populations of coral-eating snails and burrowing sea urchins, which accelerate reef erosion.[28]

The fate of the landlocked Black Sea offers an ominous warning about the dangers of bioinvasions. Weakened by a deadly combination of overfishing, coastal habitat degradation, and high levels of agricultural and industrial pollution, the sea's life has been devastated by a jellyfish-like organism, an exotic species probably introduced by a ship's ballast water in the 1980s. The Atlantic comb jelly, *Mnemiopsis ledyi*, which has no natural predator in the Azov and Black seas, feeds on fish eggs, larvae, and other zooplankton. Thanks to these invaders' taste for the young and the fact that they may be outcompeting fish larvae for food, an estimated 85 percent of the marine species in the Black Sea—including a majority of commercial fish stocks—have gone extinct in the past 20 years.[29]

Not all fisheries are overexploited and in decline. Some are not yet fished out, such as the squid and hake fisheries of the Patagonian shelf in the southern Atlantic. Populations of drum, croaker, and skipjack tuna in the Western Indian Ocean and skipjack and yellowfin tuna and Indian mackerel off the coast of Micronesia are still fairly healthy. An estimated 40 percent of world fish stocks are not yet in danger of collapse. However, none remain untouched, and because not all fish can be fully exploited at the same time, maximizing the catches of underutilized species may undercut the food source of other fish and cut into greater overall catches.[30]

It is probably not too late to avert complete collapse of the most valuable fish stocks. Some have made a comeback thanks in part to concerted management efforts ranging from quotas, fishing bans, and complete moratoria to restoration of wetlands and other habitats that function as nursery areas for larvae. Stocks of Arctic cod in the Barents Sea, striped bass in the Atlantic, herring in Norwegian waters and the Georges Bank, mackerel in the northwest Atlantic, and sardines in the U.S. Pacific, among others, have made strong recoveries. If marine ecosystems worldwide were allowed to rebuild their "original structures and biomasses, [they] could generate, on a sustainable basis, higher and certainly more valuable catches than are now taken from depleted stocks," according to fisheries scientists Daniel Pauly and Villy Christensen, of the University of British Columbia and the International Center for Living Aquatic Resources Management (ICLARM), respectively.[31]

Current marine fishing trends are unsustainable. Until fishers and decisionmakers appreciate the wisdom of Baird's simple statement—that a healthy ecosystem makes for healthy fish—efforts to restore fisheries will be in vain. Some appreciation of the complex ecological dynamics of fish populations and—given the uncertainty of this variable and fluctuating resource—built-in room for error will be essential if current levels of fishing are to be sustained—let alone increased.[32]

## *Altered Fish Habitats and Dangerous Fishing Habits*

As the United Nations was announcing the Year of the Ocean in January 1998, more than 1,600 marine scientists, fishery biologists, and oceanographers from across the globe issued a joint statement titled "Troubled Waters" that echoed Baird's warning from more than a century earlier. These experts identified many of the key issues underlying the current global fishing crisis, including pollution, habitat degradation, wasteful and destructive fishing practices, and climate change. One marine scientist summed up the current state of affairs simply as: "Too much is taken from the sea and too much is put into it."[33]

Much of what we are putting in—everything from oil that leaks into storm sewers to fertilizer that runs off farmland into waterways—ends up in the coastal zone. This region is characterized by some of the highest rates of biological productivity—and the richest fishing grounds—on the planet. An estimated 80–90 percent of the global commercial catch is hauled onto boats within 320 kilometers of the shoreline—in coastal areas, coral reefs, continental shelves, and the fertile upwelling areas where cold, nutrient-rich deep-water currents meet surface waters.[34]

Many fish species depend on coastal areas for the most vulnerable stages of life. An estimated 90 percent of commercial fish in the Bay of Bengal, for example, rely on healthy mangroves as a nursery for their young, while in East Africa and Sri Lanka, 95 percent of shrimp and marine fish live out their entire existence in coastal areas. Once abundant along the U.S. Atlantic coast, stocks of menhaden, a fish closely related to pilchard and herring that depends on wetlands for nursery habitat as well as for food, have declined by 26 percent in 10 years, in part due to the loss of coastal wetlands.[35]

Besides storing and cycling much-needed nutrients, coastal areas also collect pollution, wastes, and nutrient-rich effluent from upstream cities, farms, and industries. Although some nutrients are necessary, having too many is harmful. A growing threat to fish in many urban coastal areas is a process known as eutrophication, whereby excessive levels of nutrients build up and essentially suffocate the marine environment. In addition, seasonal algal blooms erupt off the coasts of China, Japan, and South Korea and in the Black and Baltic seas, often harboring toxic phytoplankton that poison and sometimes kill fish and shellfish. Even nontoxic blooms block sunlight, absorb dissolved oxygen, and disrupt food-web dynamics, thereby robbing marine organisms of needed food. Because of agricultural runoff in the Mississippi River basin, the Gulf of Mexico has a biological "dead zone" about twice the size of Puerto Rico that has in effect starved bottom-dwelling marine organisms and forced fishers further offshore in search of catches.[36]

Given an ever-growing human populace that gravitates more and more to coastal areas, the health of these areas is likely to worsen. Already, some 2.5 billion people (nearly 40 percent of the world's population) live within 100 kilometers of a coastline; two thirds of the world's largest cities are coastal, including Bangkok, Caracas, Jakarta, Lagos, Los Angeles, and Seoul. In the next 30 years, more than 6.3 billion people are expected to make their homes in these densely populated corridors, further stressing the seams between land and sea and crippling marine fisheries.[37]

The source of damage to valuable underwater habitats does not stop at the shoreline. Fishing gear and methods can also cause direct harm to the marine environment by damaging habitat areas, reducing cover from predators, depleting food supplies, and lowering local biological diversity. Some areas of the world's oceans are fished more than others, and therefore take a harder hit from gear. A third of the North Sea is intensively harvested each year, for instance, while an 800-square-kilometer bay in northern France is dredged—virtually scraped—seven times annually for scallops. The 40,000-square-kilometer area of gravel and marine sediments on the Georges Bank off New England was trawled with huge nets three to four times a year between 1984 and 1990, until it looked like a "parking lot," according to one researcher. By recent estimates, all the ocean's continental shelves are trawled at least once every two years, with some areas impacted several times a season.[38]

In tropical reefs, a growing threat to marine species and their habitat is the use of cyanide poison. Fishers dive down and squirt just enough sodium cyanide at the reef to stun fish, making it easy to trap them alive. Though it involves too little poison to harm people who later eat the fish, over time this practice can kill most reef organisms and convert a productive reef community into a graveyard. Divers use an estimated 400 kilograms of cyanide a year to capture valuable grouper, humphead, wrasse, Napolean wrasse, rock cod, coral trout, and other tropical species to feed a growing demand for live-food fish. Based in Hong Kong, the live-food-fish trade is valued at $1 billion a year, while the catches of ornamental species for aquariums are worth more than $200 million annually. Live fish can earn fishers 400–800 percent more income than the same species dead. Fueled by high profits, cyanide fishing is expected to drive these species to collapse in the Philippines, Indonesia, and Maldives within a few years, after which this practice will likely be spread to Papua New Guinea and the Pacific Islands. Already, cyanide fishing is reported from Fiji and Solomon Islands to Tanzania and Eritrea.[39]

**An ever present danger to marine species is the capture of innocent bystanders.**

An ever present danger to marine species worldwide is the capture of innocent bystanders. In the process of taking fish, unwanted species are often brought on board and then thrown back into the sea, often dead or dying. Few of them survive the process of being yanked out of their habitat and then later dumped overboard. Known as discards, these unwanted fish are wasted either because they are undersized or a nonmarketable sex or species, or because a fisher does not have a permit to catch them. FAO estimates that discards of fish alone—not counting marine mammals, seabirds, and turtles—total 20 million tons, equivalent to one fourth of annual marine catch. Global bycatch—the sum of discards and unintentionally caught species that are retained—was estimated at more than 28 million tons in 1994.[40]

To a large extent, bycatch is associated with industrialized fishers, who use indiscriminate gear to catch as much as they can, but generally keep only the fish they are legally permitted to catch or those that make money. A standard commercial trawl net designed for groundfish, for example, can catch high quantities of shrimp, crabs, and even seals. Because fish processors and commercial markets accept only certain types of fish, the nontarget species are less valuable to industrial fishers. Although markets and government regulations encourage fishers to be particular about what they keep, their gear is by no means selective.[41]

The trawl fishery of the Northwest Pacific—which targets pollock, yellowfin sole, Pacific cod, crab, and mackerel—produces the largest quantity of bycatch in the world. Of the 27 million tons of fish caught in the Northwest Pacific region in 1995, some 9 million tons were dumped back in the sea. Discards of crabs and their prey from the groundfish fishery caused losses of more than $50 million in the Bering Sea crab fishery in 1991 alone. Losses of all species in the Bering Sea and Gulf of Alaska combined are estimated at more than $250 million annually.[42]

Because of small-mesh nets and the high species diversity in tropical waters, shrimp trawling also produces large quantities of bycatch. Off the Brazilian coast, discards in the shrimp fisheries are comparable in volume to the total landed catch. Worldwide, for every kilogram of shrimp caught, at least 5 kilograms of other species are discarded; in some regions, the ratio is 1 to 15.[43]

While the threats destructive fishing gear and accelerating coastal development pose to fish are clearly caused by people, other factors—such as changing weather patterns and variable oceanographic conditions—seem far removed from direct human involvement. But there are indications that human-induced climate change is already influencing the marine food web, and there is little doubt that climate change will exacerbate existing pressures on marine resources. Fish are especially vulnerable to temperature change as they cannot regulate their internal temperature. Warming waters and changes in ocean salinity and circulation can redistribute important food supplies and alter localized biological productivity, sometimes triggering abrupt and massive shifts in the abundance and range of fish species.[44]

During autumn and winter 1997, El Niño warmed waters off the coast of California to their highest temperature in 40 years of record-keeping. As a result, tropical fish such as mahi mahi, albacore tuna, swordfish, and sardines had to swim northward up the coast of California and along Oregon and Washington in search of food. If waters off the western coast of Canada warm by 2 degrees Celsius, as they are expected to by 2070, Pacific salmon may have to swim all the way to the Bering Sea to find adequate food supplies.[45]

In March 1995, scientists from Scripps Institute of Oceanography reported that the ocean temperature off the coast of San Diego, California, had warmed by 0.8 degrees Celsius during the previous 40 years. As a result, concentrations of zooplankton, microscopic animals that feed on the base of the marine food chain, had declined by 70 percent in the preceding two decades. This loss severely affected the food supply of several commercial stocks, including sardines, anchovy, hake, jack mackerel, and Pacific mackerel. Indeed, commercial anchovy fishing in the area has completely collapsed and other commercial fish species have declined by as much as 40 percent since the early 1970s, in part due to changes in ocean temperature. Similarly, the population of sooty shearwaters, a native seabird, declined by 90 percent between 1987 and 1994 because they could no longer find enough food to eat.[46]

If current predictions by the Intergovernmental Panel on Climate Change (IPCC) prove correct, in the next 50–100 years climate change could have a greater impact on the health of world fisheries than overfishing itself. Climate change will likely exacerbate the effects of pollution, habitat degradation, and ultraviolet radiation. Expected rises in sea level may effectively drown estuaries, thus destroying habitats for fish and their prey. Some fish species may migrate toward the cooler polar areas in response to rising sea temperatures at lower latitudes, while others will grow faster in warmer water, and outcompete other species. Waters warmer than 19 degrees Celsius, for example, favor redside shiners over the more commercially valuable juvenile steelhead trout.[47]

A more pernicious effect of climate change may result from the damage caused by increased ultraviolet (UV-B) radiation, which can weaken each link in the marine food chain. The thinning ozone layer above Antarctica has allowed more UV-B radiation to penetrate the waters, depressing photosynthesis and the growth of phytoplankton and macroalgae under the polar icecaps by as much as 5 percent. The effects are not limited to the base of the food chain: a 1997 study revealed an increased incidence of DNA lesions in eggs and larvae of Antarctic icefish likely due to increased intensity of ultraviolet light. Similarly, solar UV-B radiation has been found to damage the larval development of species such as crab, shrimp, and some fish. By striking aquatic species during their most vulnerable stages of life, and reducing their food supply at the same time, UV-B could have devastating impacts on world fisheries production. In a 1988 study, marine scientists demonstrated that a 16 percent loss in the ozone layer would cause a 5 percent decline in phytoplankton, which could lead to the loss of some 7 million tons of fish per year.[48]

If depleted stocks were rehabilitated, levels of bycatch were reduced, and selected species were carefully brought to peak production, global marine catches could reach 125 million tons—a boost of 40 million tons from today's levels—according to FAO estimates. However, reaching that potential will depend largely on the extent to which fishers turn away from destructive methods and refrain from stepping up their efforts in the face of diminishing catches.[49]

## *Fleeting Returns*

A primary cause of overfishing is the fact that world fishing fleets are simply too large for the available resources. Undergirded by a system of government bailouts that has propped it up for the last several decades, today's fishing industry is dysfunctional and massively overcapitalized, with too many big boats and too many fishers taking too much from the sea. Although becoming more evident among smaller-scale outfits, the problem is far worse in the commercial sector of top fishing nations, which field the world's largest vessels and the greatest share of the fishing technology.[50]

Several powerful forces are driving fishers down the path of self-destruction by encouraging overcapacity. They include the open access nature of fishing—which draws people into the industry well after profits and catches begin falling; widespread technological change and modern equipment that have increased the efficiency of marine fishing and replaced workers; and national claims to fishing grounds combined with economic development policies that have promoted growth in all sectors of the industry without providing the means to ensure conservation and protection of the resource.[51]

In open access fisheries, resources are owned by no one individual. Fishers tend to have little incentive to practice conservation, for they know that if they do not catch the available fish, someone else probably will, and without limits in place, fishers try to catch as many fish as they possibly can. Then, in response to declining yields, most fishers tend to overinvest, often with undesirable consequences. Although overexploitation is by no means the inevitable outcome, it is the usual one.[52]

For large fish populations harvested with simple technology and by relatively few fishers, open access seldom causes serious problems. In fact, in an underexploited fishery, fishers initially enjoy large returns. But strong earnings usually attract more people to the fishery, which in turn prompts those already there to invest in bigger boats and to exert more effort to maintain or increase their share of the catch. The first fishers sustain the lowest costs because fish are easier to find and more plentiful. But by the time many people are pursuing the resource, the fish are less concentrated, less plentiful, and harder to find, so it costs more to harvest them.[53]

The conditions of open access can lead to the scenario described by Garrett Hardin in his seminal 1968 *Science* magazine article, "The Tragedy of the Commons": an unrestricted number of users, unfettered by any limits on their access, extract an increasing share of a resource until natural resources are severely depleted, sometimes to the point of no return. In fisheries, as in other renewable resources that are open to the public, the tragedy stems from the fact that a large portion of the commons is relatively unmanaged, a condition that invites careless and shortsighted use rather than collective restraint.[54]

Generally, the harder fishers try, the more fish they will catch—but only up to a certain point, known as maximum sustainable yield (MSY). MSY represents the largest amount of fish that can be extracted at a certain level of fishing pressure while still ensuring an adequate bounty the following year. Anything beyond this point is considered overfishing. Although precise in description, MSY is difficult to pinpoint for one species, let alone for several interrelated species because of the large number of variables that influence each fish population. The process of determining sustainable yield of this fluctuating resource is often more art than science.[55]

How quickly the point of overfishing is reached depends on the health of the fish populations, the quality or reliability of data and methods used to determine MSY—which serves as the basis of annual catch limits—and whether fishers respond to declining biological returns by temporarily holding back. When fishing increases too much or too fast, or the resource itself is declining due to natural causes, the risks of overfishing are higher. The conditions of open access tend to hasten the day of overshooting maximum sustainable yield by making it cheap for individual fishers to exploit the resource and letting society at large bear the costs. By the time multiple users go beyond MSY, it is often too late. Since most fishers are already too heavily overcapitalized for their average catch rates, only the most efficient outfits can break even. In order to pay their bills, fishers have to keep working even when they know their efforts will ultimately backfire. Some fishers will go out of business when an economically inefficient fishery becomes too crowded, but with the help of government aid or subsidies, too many others will stay put.[56]

Today's overcapacity problems have their roots in the uninhibited growth and rapid modernization in the industry over the past 50 years. Building on technological developments during World War II, fishing underwent dramatic changes. Many industrial nations converted their navy vessels to fishing boats and made use of electronic navigation systems, ship-based surveillance technology, and sonars for their commercial fishing fleets. European nations, Japan, and Russia built large fishing vessels that could travel to distant waters, while the World Bank and other multilateral aid agencies focused on developing fisheries to reduce poverty, encourage overall economic development, and "improve the efficiency" of traditional fishers.[57]

**Several powerful forces are driving fishers down the path of self-destruction.**

In the early 1950s, workers in European shipyards built huge, mechanized vessels that could catch up to 500 tons of fish a day. Known as factory trawlers for the enormous trawl nets that are hauled up the ship's stern to onboard processing and freezing facilities, these floating factories allowed fishers to catch and fully process large quantities of fish. Previously, ships had to return to port before the catch spoiled. But once fishers became able to process, freeze, and package fish at sea, they were free to roam the globe in search of profits.[58]

Factory trawlers concentrated on temperate and cold-water offshore fisheries, which typically have high concentrations of single or similar species. With improved reconnaissance and communications systems, fleets of these massive vessels were able to zero in on large schools of fish, harvest the entire concentration, and then locate and converge on the next large congregation of fish. By the mid-1970s, factory trawlers from northern Europe, Russia, and Japan were removing a large share of the world's pelagic fish from the high seas, as well as a significant portion of demersal species in the North Atlantic and Northwest Pacific oceans.[59]

From their first use, factory trawlers raised the potential for profits—and overfishing. In the highly productive Georges Bank waters off the coast of New England, for example, nearly 1,000 huge trawlers from foreign countries captured more than 2 million tons of groundfish in 1974—about 10 times as much as was caught by U.S. fishers that year.[60]

A backlash against these mechanical predators heated up soon after their initial launch. Many coastal states reacted to the intrusion of foreign vessels by extending their national authority further offshore. In 1945, the United States was the first country to extend its control from the traditional 12-mile territorial zone to contiguous high seas. Under the Truman Proclamation, U.S. officials justified the move as a way to better protect fisheries, establish conservation zones, and exploit seabed minerals of the continental shelf. Many fishing-dependent countries soon followed suit, triggering a global "sea grab." Within a decade, several Latin American countries, including Argentina, Peru, Chile, and Honduras, extended their jurisdiction to 200 nautical miles to protect their fisheries from outside intrusions.[61]

What began as an isolated trend in the 1950s and 1960s quickly grew into a global phenomenon. By 1973, nearly 35 percent of the ocean's area—equal to the Earth's entire land mass—was claimed by coastal states, many of them developing countries. These claims led to the third and final U.N. Convention on the Law of the Sea (UNCLOS) in 1982, which did not formally enter into force until the required 60 countries had ratified it by November 1994. Under UNCLOS, coastal nations were granted rights to use and develop fisheries within a 200-nautical-mile exclusive economic zone (EEZ). With the privilege of harvesting the marine resources came an obligation to protect and conserve fish stocks. The passage of this convention marked the end of an era: freedom of the seas no longer existed in the 200 nautical miles closest to shore, though the high seas remained open.[62]

Around the time nations were extending their claims to fishing grounds, international development agencies boosted their lending to expand and improve the fishing sector in developing countries. The World Bank, the Asian Development Bank, and the aid agencies of Europe, the United States, and Japan provided millions of dollars in financial and technical assistance to governments in coastal developing countries. Specifically, programs were aimed at modernizing fishing technology and vessels, building shore-based support facilities and related infrastructure, and developing market-oriented aquaculture projects. Fishers were encouraged to catch as much as possible, and to export a portion of their haul to industrial countries to help their own nations generate foreign exchange.[63]

As a result of EEZ claims and outside financial assistance, developing countries contributed a growing proportion of world fishing capacity. Between 1970 and 1989, these countries' share of the world fleet increased from 27 to 58 percent. By 1994, an estimated 43 percent of the world's fishing capacity was concentrated in Asia.[64]

By redistributing control away from predominantly northern distant-water fishing nations to coastal countries worldwide, the Law of the Sea essentially democratized world fishery resources. Indeed, this reallocation is one of the lasting legacies of UNCLOS. In the early 1950s, about 80 percent of the world's fish catch was taken by industrial countries. Forty years later, 64 percent of the catch was in the hands of developing countries. By 1994, developing and newly industrializing countries accounted for seven of the top 10 fishing nations: China, Peru, Chile, Russian Federation, Thailand, Indonesia, and Republic of Korea.[65]

Although it did provide benefits, the process of nationalizing waters conflicted with the multinational reality created by long-range fishing fleets, and had a number of undesirable consequences. The Law of the Sea left the tricky part of fisheries management—how to divide up the catch among boats and people—up to individual nations to hash out in the field. Many nations repeated the mistakes of the past by overfishing their own and other nations' waters—a pattern that has long existed in the North. Coastal state governance was effective at encouraging growth and development in world fisheries, but it had the unintended effect of accelerating the pace of overfishing.

During the 1970s and 1980s, the gross registered tonnage of world fleets, a measurement of volume, increased by 90 percent, while the technical capabilities of the world fleet as a whole increased more than three times as fast, by 330 percent, signifying a massive escalation of fishing power and effort. Despite the investments and improvements in fishing technology and harvesting capacity and the growth in world fish catches, landings per gross registered ton (catch rate) declined by 62 percent overall during these two decades. (See Figure 3.) Large boats were catching less for the same amount of effort—a direct consequence of overcapitalization.[66]

By the late 1980s, the world's large-scale fishing fleet had exceeded the maximum sustainable yield of all the world's commercial fish stocks by 30 percent. If fishers wanted to maximize economic efficiency, they would have to reduce capacity between 25 and 53 percent, depending on price increases or cost reductions. Their other options were equally difficult: lower operating costs by 43 percent, or raise consumer prices by 71 percent, according to FAO analysts.[67]

Since 1989, the situation has grown even more precarious. Additions to the world fleet still exceed deletions and technical capacity continues to mount. Between 1992 and 1997, the world's fleet increased by 3 percent in tonnage and 22 percent in technological capability. Current fleets now have at least 50 percent more capacity than they need for world fishery resources. For some individual fisheries, the problem of over-investment is even worse.[68]

The value of the global catch for the most part is declining or stagnant, even though catches are still increasing in some areas, and prices of some individual species have

**FIGURE 3**

**Global Fleet Capacity and Catch Rate, 1970–89**

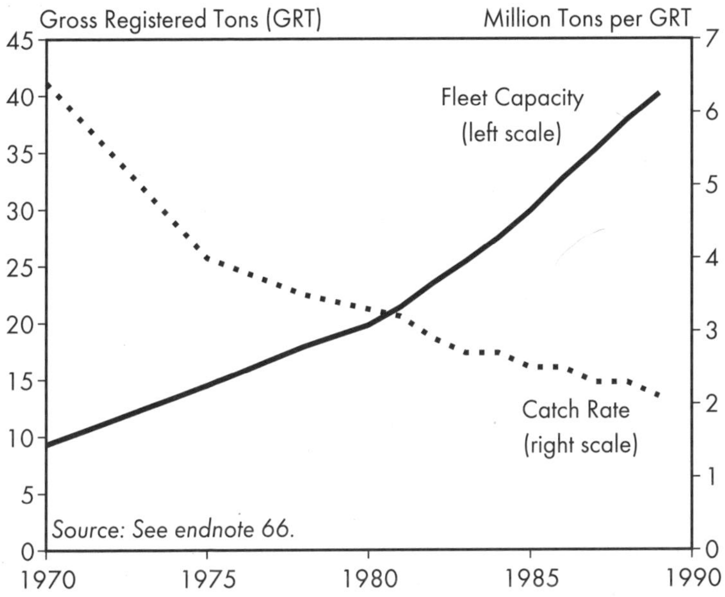


risen considerably. In 1989, fishers in all sectors were earning less than half of what they had collected 20 years earlier per unit of fishing power, despite having invested in improved fleets. The reasons for this slippage include changing catch composition, lower-priced imports, and, for some species, cheaper farmed versions.[69]

Rising costs and falling revenues have made the industry financially vulnerable. In 1989, worldwide losses from overfishing amounted to roughly $54 billion, based on estimated global fishing costs of $124 billion and fishing revenues of $70 billion. Even where fishers continue to make money, they run the risk of losing it all in a couple of years of poor catches. Finding themselves caught in an economic trap of mounting debt and declining yields, fishers have pressured governments to keep excessive fishing quotas intact.[70]

One key reason that so many fishers remain afloat is that governments bear a growing share of the losses—

through tax incentives, low-interest loans, and direct subsidies. For every dollar earned from fishing in the late 1980s, governments, taxpayers, and fishers spent $1.77, a debt that quickly translated into shrinking profits, unpaid loans, and job layoffs, while at the same time stimulating greater overexploitation and driving governments further into the red.[71]

Despite the losses in the late 1980s, many governments today continue to give fishers immense amounts of subsidies. Most of this money actually bolsters fishing capacity and upgrades existing boats, thus encouraging fishers to try to catch even more fish. Using data from the few governments that keep track of these expenditures—China, the EU, Japan, Norway, Russia, and the United States—Matteo Milazzo, a researcher with the U.S. National Marine Fisheries Service, estimated that global fishing subsidies in 1995 totaled $14 to $20 billion. Between $3.0 and $3.5 billion were budgeted specifically for domestic fishing subsidies, plus $1 billion for buying access rights in foreign waters. Tax breaks and lending totaling $3 billion acted as subsidies for buying fishing boats and gear. An additional $7 to $11 billion came from unbudgeted subsidies and low-interest loans and tax preferences for shipbuilding, harbor development, and related infrastructure projects. Based on these data, 20 to 25 percent of current global fishing revenues come from subsidies. Milazzo argues, however, that these estimates are "cautious and truly conservative." Subsidies on the order of 25 to 40 percent are more likely.[72]

Primarily bestowed by industrial countries on failing fleets that continue to operate under open access conditions, subsidies encourage the tragic phenomenon of fishers participating in the downfall of their own industry. While subsidies may help to cushion the economic impact of declining fisheries in the short term by shielding fishers from the bracing effect of the market, over the long run, they encourage recipients to remain in the industry and to continue overextending themselves financially, thus further straining the resource base.[73]

Perhaps the best way to fathom the full cost of subsidizing world fishing fleets is to appreciate what overcapacity and overfishing cost in lost revenues and employment. In the United States alone, an estimated $8 billion and 300,000 fishing-related jobs could be gained from better management and protection of fish stocks, according to researchers from the National Marine Fisheries Service. Worldwide, fishers forfeit profits on the order of $25 to $30 billion and marine catches of nearly 20 million tons because fisheries are not managed properly.[74]

Thus, the expansion of the world's fishing fleet and maintenance of marine catches in recent years have come at a steep price for governments, fishers, and the resource itself. Clearly, the fishing sector must scale down its efforts, but how to shape this change—to determine who will exit, who will remain, and at what cost—presents grave challenges to policymakers.

## *Fishing for Whom? By Whom?*

In the village of M'Bour, Senegal, few fish are wasted or discarded. The highest-quality fish are diverted to export markets while the lower-quality ones are smoked, dried, salted, and sold in local markets. These cheaper, preserved fish are the key to food security in Senegal, a nation of 8.8 million people who rely on fish for 75 percent of their animal protein. With increasing pressures on land-based food systems, the need to preserve adequate supplies of fish for local consumption is greater than ever. Local fishers also face increasing competition from outsiders, a threat that is becoming more significant as more people from other countries come to depend on the fish from Senegal's waters.[75]

Some 20 years ago, Senegal, like several of its neighbors in West Africa, received funding from the World Bank to upgrade its fishing capacity and purchase new fishing vessels. Today, the gutted hulls of these ships lie abandoned on beaches—there are no parts or equipment to fix them, and fuel is scarce. Standing beside the skeletal remains, one can occasionally pick out the latest incarnation of outside "assistance," offshore on the not-so-distant horizon: European trawlers allowed into West African waters through access agreements with national governments.[76]

During the latest round of negotiations with the government of Senegal in 1996, the European Union secured rights for the first time to valuable coastal species, including sardines and mackerel, the very species that people in Senegal eat. Fearing the loss of their food—and their way of life—the fishers of M'Bour and other villages have appealed to their government to ban fishing in key spawning areas.[77]

Worldwide, rising demand for food has a profound impact on who fishes and who eats fish. In undermining marine and coastal resources, overfishing and overcapacity contribute to a second tragedy less appreciated than that of the commons: the tragedy of the "commoners." A highly profitable global seafood trade is driving overfishing, which in turn stokes a vicious cycle of economic hardship and social disruption for great numbers of smaller-scale fishers. Meanwhile, ownership and the benefits of resource exploitation increasingly concentrate in the hands of the few. Although fish is traditionally known as the "poor man's protein" and fishing is seen as the employer of last resort, as fish become increasingly scarce and more expensive, they are eluding the grasp of the poor.[78]

As in Senegal, low-quality fish in many developing countries remain the least-expensive option to meet minimal nutritional needs. Throughout Sub-Saharan Africa, smoked and dried fish comprise more than half of the fish consumed, especially by the poor. Fish is also a staple food in Bangladesh, where the old proverb "fish and rice make a Bengali" still holds true today. Although only 6 percent of the typical Bengali diet is animal meat, 50 percent of that comes primarily from small (less than 25 centimeters long), indigenous fish that are caught in floodplains and inland waters. People eat these small fish whole—including bones, organs, and skin—thus obtaining 90 percent of the vitamin

A, and some 15 to 30 percent of the calcium and iron that they need.[79]

People get 16 percent of their animal protein from fish worldwide. But the inhabitants of developing countries generally rely on fish for a much larger share of their animal protein than people in industrial countries do. People in a handful of countries—North and South Korea, Ghana, Indonesia, Congo, Japan, Malawi, and the Philippines—depend on fish for more than half of their animal protein needs. Nearly 1 billion people worldwide, predominantly those living along coasts and rivers in Asia, rely on fish for 30 percent of their animal protein needs.[80]

**Rising demand for food has a profound impact on who fishes and who eats fish.**

Available per capita fish supplies vary considerably, depending on the size of the human population, the volume of fish produced and traded, and the quantity of fish that is used for non-food purposes. The total food supply per country and by region is calculated as the sum of total production (catches and culture), minus non-food uses, plus imports, minus exports. About a third of world fish production, 30 million tons, are dedicated to nonfood uses, such as fishmeal and oils, which left just 71 million tons available for direct human consumption in 1992. That same year, available fish supplies per person stood at 13 kilograms worldwide. Within Asia, 23.5 kilograms of fish per person were consumed in Southeast Asia, compared to 4.4 kilograms in South Asia, and 12.9 kilograms in China alone. In sharp contrast, Japanese consumers enjoyed the highest rate of per capita fish consumption in the world—67 kilograms.[81]

At an average of 27.9 kilograms per person per year, people in industrial countries consume three times as much fish as do people in the developing world, who consume an average of 9.2 kilograms. Yet people in developing countries rely on fish for a much larger portion of their animal protein than people in industrial countries do. (See Figure 4.) Residents of industrial countries, who represent less than

**FIGURE 4**

**Available Per Capita Fish Supply and Share of Total Animal Protein, by Region, 1992**

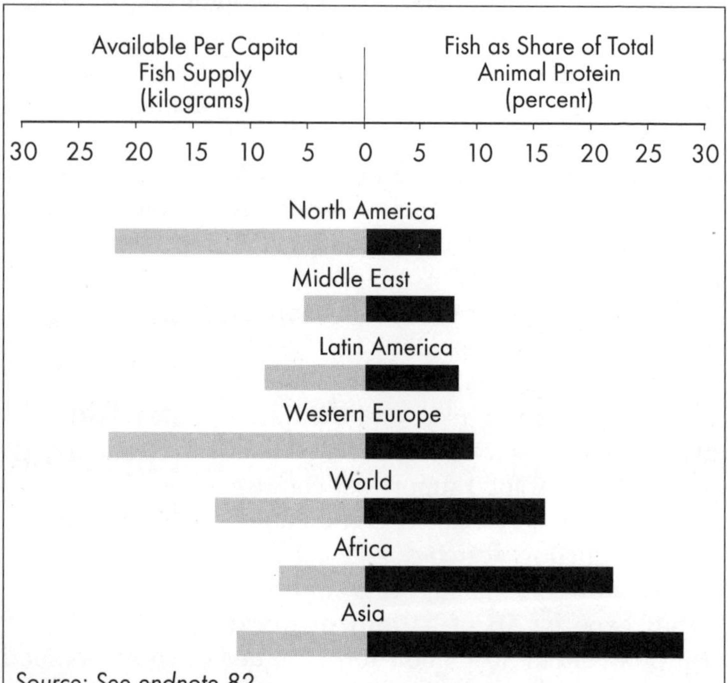


*Source: See endnote 82.*

one fourth of the world's population, consume 40 percent of the fish, while the three fourths of the world's population who live in developing countries consume only 60 percent. The 9.4 million tons of fish used for non-food purposes in industrial countries in 1991 exceeded the total supply of fish for human consumption in Latin America and Africa combined.[82]

Despite historical growth in world fish production, per capita available supplies are likely to decline in the near future. In some developing countries, this turnabout has already taken place. Since 1961, daily supplies per person have increased by more than 70 percent in industrial countries and by 50 percent in all developing countries. During the 1980s, however, per capita fish supplies showed a slight decline in Africa and Latin America. With most of the world's future population growth expected to occur in developing countries—the source of half of global fish supplies—the declines in per capita supplies will soon be felt globally.[83]

These declines are in part due to the fact that nearly half of the fish caught today is traded between nations, compared with an estimated 32 percent in 1980. The value of the global fish exports has increased more than fourfold, from $11 billion in 1970 to $52 billion in 1995 (in constant 1995 dollars). (See Table 1.) Consumers in industrial countries account for 83 percent of world fish imports by value but just 61 percent of volume. Industrial countries import a large share of the world's high-priced species such as tuna, shrimp, squid, and salmon, as well as large quantities of fishmeal for animal feed and fertilizer. Worldwide, more than 70 percent of all frozen and canned tuna, and 77 percent of all frozen shrimp come from developing countries.[84]

To feed industrial countries' growing appetite for seafood, several developing countries are now solid net exporters. (See Table 2.) With profits from a fairly narrow product range, fishers in Thailand, China, Chile, and Indonesia generate the lion's share of the net trade balance, and their export business has soared by more than 1,400 percent since 1970. Thailand's rapid ascent to number one position among world fish exporters during the past decade, for example, has been achieved through the development of a canned tuna industry and a 1,000-fold increase in giant tiger prawn culture.[85]

Although dominated by flows from South to North, fisheries trade is not exclusively one-way. India, Indonesia, Malaysia, the Philippines, and Thailand import fishmeal from Chile and Peru to feed their export-oriented shrimp farms. Likewise, processing plants and canneries in Thailand and the Philippines import sardines and mackerel from Latin America and frozen tuna from the United States, with the end product usually shipped back to the West.[86]

In many ways, growth in fish trade is a double-edged

**TABLE 1**

## Fish Imports and Exports, by Quantity and Value, 1970–95

| | Quantity | | | Value | | |
|---|---|---|---|---|---|---|
| | 1970 | 1995 | 1970-95 | 1970 | 1995 | 1970-95 |
| | (mill. tons) | | (percent change) | (1995 $ billion) | | (percent change) |
| **Imports** | | | | | | |
| Developing Countries | 1.8 | 8.2 | 356 | 2.3 | 9.8 | 326 |
| Industrial Countries | 5.6 | 13.0 | 132 | 10.7 | 46.2 | 332 |
| World Total[1] | 7.5 | 21.2 | 183 | 13.0 | 56.0 | 331 |
| **Exports** | | | | | | |
| Developing Countries | 3.8 | 12.6 | 232 | 4.1 | 29.1 | 610 |
| Industrial Countries | 3.6 | 9.2 | 156 | 6.8 | 22.9 | 237 |
| World Total[1] | 7.4 | 21.7 | 193 | 11.0 | 52.0 | 373 |

[1]Some numbers do not add up due to rounding. Data include all types of fish, shellfish, oils, and fishmeal.
*Source:* See endnote 84.

sword for developing countries. Exporting high-end fish brings in foreign exchange for economies desperately in need of cash to pay off debts. In 1995, developing nations earned $19 billion from fisheries trade. Increased income can certainly contribute to better nutrition if combined with education, family planning, and a whole range of social development programs. Trade also provides a safety net for people by allowing for imports of cheaper fish, with large net nutritional gains.[87]

Yet while offering these benefits, trade can also pose a serious threat to the growing need for food security, especially among the less affluent. While shortfalls in fish supplies and price hikes may be little more than an annoyance

TABLE 2

## Top 10 Fish Exporters, by Value, 1970–95

| Country | 1970 Exports (1995 $ mill.) | 1995 Exports ($ mill.) | Share of 1995 World Total (percent) | Increase since 1970 (percent) |
|---|---|---|---|---|
| Thailand | 69 | 4,449 | 8.6 | 6,348 |
| United States | 439 | 3,383 | 6.5 | 671 |
| Norway | 1,021 | 3,122 | 6.0 | 206 |
| China | n/a | 2,854 | 5.5 | n/a |
| Denmark | 651 | 2,459 | 4.7 | 278 |
| Taiwan | n/a | 2,328 | 4.5 | n/a |
| Canada | 1,011 | 2,314 | 4.5 | 129 |
| Chile | 107 | 1,704 | 3.5 | 1,493 |
| Indonesia | 21 | 1,666 | 3.2 | 7,833 |
| Russia[1] | 355 | 1,628 | 3.1 | 358 |
| World | 11,576 | 52,034 | | |

[1]Figure for "Russia" is actually for Soviet Union. Data for Russia alone is not available.
*Source:* See endnote 85.

to better-off consumers in industrial countries, people in the middle class in India and the Philippines, for example, already feel that they can no longer afford to eat fish as part of their regular diet. In the past five years, world seafood prices have risen by 4 percent annually. While the trends in fish prices are complicated by growing supplies of cultured species, which can lower prices, trade does tend to exacerbate short-term declines in supply and thereby to trigger higher local prices. Once it was known as the poor people's food, but today, "when fish supplies deteriorate, fish tends to disappear first from the plates of the poor," according to George Kent, a political science professor at the University of Hawaii.[88]

In recent years, per capita fish supplies available in Laos and Cambodia have fallen short of meeting subsistence needs. Despite the rich fishery resources in these two countries, many poorer people now suffer from malnutrition and protein deficiencies. A similar pattern may soon occur in Malaysia and Suriname, where exports have increased but total fish production has dropped. India and Mexico, however, have been able to increase production fast enough to keep pace with rising exports so that domestic supplies have not yet declined, but prices are on the rise. As a growing portion of domestic production feeds consumers abroad, inevitably the net supply for domestic consumption drops.[89]

Traditionally, fishers have been the employer of last resort. But in some coastal areas they are the employer of only resort. According to the most recent FAO statistics, more than 21 million people worldwide are fishers (compared to 9 million people involved in aquaculture). Almost 90 percent of all fishers are small-scale operators and 95 percent live in developing countries. Coastal fisheries in South and Southeast Asia alone employ about 6.7 million people, and when employment of related industries such as processing and canning are taken into account, the number rises to 15 million. In M'Bour, Senegal, the fishery supports a complex local network of women who handle and prepare the fish, people who transport the catch by horse or truck, stall owners, and of course, the fishers themselves. More than 200 million people around the world depend on fishing for their income. Although fisheries may not add substantially to GNP, their contribution to employment and food security is significant. More than half the fish eaten today come from inshore and coastal areas that are dominated by more than 19 million small- and medium-scale fishers who are officially counted in FAO statistics. (A large number of subsistence, part-time, and seasonal fishers are not included in these estimates.)[90]

Depending on the scale of operation, fishing offers vastly different benefits to the people involved and to the resource itself. Small-scale fishers, for example, garner lower wages and income per person but produce a higher share of fish for direct human consumption. Smaller-scale fishers also tend to keep everything they haul up in their nets and generally cause less damage to the resource base than large-scale, commercial fishers do. In contrast, a substantial portion of fish caught by large-scale, industrial fishers is processed into fishmeal—which goes to feed livestock, poultry, pets, and farm-raised fish, or into cooking oil, catfood, and fertilizers. The catch for nonfood uses is on par with what commercial fishers contribute to direct human consumption. And the quantity of wasted bycatch in their operations is substantial.[91]

In developing countries, as growing populations crowd onto increasingly degraded interior land, large numbers of people are moving to the coasts. With no discouragement from governments, people—including Peruvian highlanders, traditional pastoralists in West Africa, and landless rice farmers in the Philippines—are now flocking to the fishery. Naturally, the entrance of nontraditional fishers into this sector has exacerbated pressures on the resource and increased tensions, particularly in small-scale fisheries. Newcomers are sometimes more willing to go to extreme measures to capture fish, to the point of degrading the resource and putting their own lives in danger with cyanide poison, for example, because they have little to lose. Their presence accelerates the downward spiral of resource degradation, but it is by no means the primary cause of the problem.[92]

Commercial vessels and foreign-owned companies working off shore bear a large share of the responsibility for overexploitation of world fishery resources. With more expensive gear needed, fisheries are increasingly dominated by corporate vessels and absentee boat-owners. Kjell Inge Rokke, a Norwegian-based conglomerate, controls almost 10 percent of the world's whitefish (cod, hake, and pollock) production, with operations concentrated in the North Pacific and Russia. Similarly, the Spanish company Pescanova accounts for 20 percent of world hake production, with ventures in South Africa, Namibia, and Mozambique. In early 1997, Yamaha Motor Company announced that it would buy fish feed from Dutch producers to sell to Japanese aquaculturists. Large multinational companies such as these have the required financial backing and political influence both to pressure their own governments to underwrite their efforts to remain financially solvent and to persuade foreign governments to give them cheap access.[93]

Boats from Europe, North America, and Northeast Asia have fished in waters of the southern hemisphere for decades. For nearly a century, the Japanese government has supported policies encouraging its fishers to work in distant waters of the South Pacific. Since the 1970s, European fleet owners—with the help of the EU—have signed access agreements with individual West African nations to fish in their waters. The process of countries declaring their EEZs under the Law of Sea changed the playing field by granting coastal countries a legal basis for excluding foreign fishers. UNCLOS also gave these nations the right to charge fees as rent for access to the resource.[94]

Today, fisheries in the South are for the most part already exploited near their limits by southern nations' own increasingly mechanized fleets and by small-scale fishers close to shore. With mounting pressures from both foreign and domestic boats, the terms of access agreements are coming under increasing scrutiny. As demonstrated in the EU agreement with Senegal, a key issue concerns the benefits countries forgo by allowing others to extract their own natural wealth—and who benefits from the price paid for access. To give an idea of what is at stake, a recent study concluded that export earnings from one domestic tuna vessel in Papua New Guinea exceeded net revenues from the licenses and access fees of 130 foreign boats.[95]

With severe overfishing in the northern hemisphere, industrial countries are now willing to pay a high price for access to southern EEZs. In 1996, the EU paid $229 million—or 43 percent of the EU's annual monies earmarked for addressing overcapitalization—for access agreements with Africa, primarily for the benefit of French, Portuguese, and Spanish fishing companies—thereby exporting the overcapacity problem from North to South. The vessel owners themselves pay only a fraction of the cost.[96]

In June 1996, the EU signed a $70 million-a-year fisheries access-for-trade agreement with Mauritania. This latest arrangement expands European rights along the 750-kilometer coastline considerably. It stipulates a 45 percent jump in the number of boats and a 140 percent increase in allowed catches as well as first-time capture rights for highly prized squid and octopus—despite the fact that Mauritania's fisheries are already fully fished and some species are overexploited. In both Mauritania and its neighbor to the south, Senegal—two countries whose populations are growing faster than 2.5 percent per year—the domestic fishing industry and local fishermen's associations protested the terms of these agreements. But their governments have largely ignored their opposition.[97]

From the point of view of African nations, debt service is the key motivation for annual renewal of access agreements. But the results of the agreements have been profoundly negative, according to Okechukwu Iheduru, an assistant professor of International Relations at Michigan State University. Pelagic fish are harvested at minimal prices by European fishing companies, and Africans are left with overfishing and, increasingly, undernourishment.[98]

With few enforceable limits in place, this transfer of fleets from European to West African waters represents an international environmental justice issue. Local fishers are rarely consulted, the foreign fishers regularly underreport catches (which form the basis for future agreements), and enforcement of the few environmental provisions that do exist is generally scant. The increase in legal foreign fishing, combined with still-rampant illegal foreign fishing, is certain to harm the resource. The EU currently subsidizes similar arrangements with 14 other nations in West and East Africa—paying less than 10 percent of the value of the catch in most cases.[99]

South Pacific island nations also have a history of granting foreign fishers access to their tuna fisheries, even though subsistence fishing supports 78 percent of the region's population. The importance of tuna resources to the world and to the South Pacific island nations themselves is enormous: half of the world's canned tuna is caught in the South Pacific; tuna constitutes 30 percent of exports from Solomon Islands and 40 percent of GDP in Kiribati. Valued at nearly $1.5 billion, about 1 million tons of skipjack, yellowfin, bigeye, and albacore tuna were caught in 1995. Nearly 5 percent of this value was paid to South Pacific nations in the form of access fees. The remaining 95 percent went to the captains, owners, and crews of fishing fleets from the United States, Taiwan, Japan, and Korea.[100]

Unlike their counterparts in West Africa, Pacific Islanders have tried to execute foreign access agreements as a regional block, through the South Pacific Fisheries Forum (SPFFA). Finding their attempts thwarted by distant-water nations—only the United States has negotiated with the SPFFA as an entity—individual SPFFA members have cooperated on setting minimum terms of access, developing a regional registry of fishing vessels, and stipulating catch-reporting requirements. Moreover, these nations cooperate in surveillance and enforcement. A foreign vessel caught fishing without a license in the EEZ of one of the member nations will find itself blacklisted throughout the region.[101]

Arguably, greater financial benefits could be secured through a restructured agreement—Japan currently pays Russia about 30 percent of value for landing salmon in Russian waters—six times more than the share they pay to Pacific Islanders for tuna access rights. In a sign of things to come, Papua New Guinea recently terminated its 40-year policy of granting tuna fishing licenses to foreign vessels. In February 1998, a new PNG-based fishing company, Sokimona, secured $600 million to construct 20 modern tuna purse seiners, onshore processing facilities, and associated support systems.[102]

In developing and industrial countries alike, local consumers, small-scale fishers, and households dependent on fishing for their food and livelihoods simply cannot compete with big factory trawlers, foreign-owned companies, and distant export markets. (This is especially true if the enormous subsidies that currently favor large-scale operations are kept intact.) Worldwide, if per capita food supplies are to remain stable, at least 91 million tons of fish will be needed by 2010 for direct human consumption—7 million tons more than were available in 1996. To meet growing demand in the future, developing countries may have to use a greater share of earnings from high-priced fish exports to import lower-cost fish from abroad. They will also need to reconsider the terms and price of access for foreign fishers. With more than one third of global fish production directed to nonfood uses, livestock and fish farmers would do well to substitute other sources of protein, such as soymeal, for fish in animal and fish feed. And governments everywhere may want to reexamine the role of aquaculture in fish supplies and then invest in types of fish farming that are beneficial both to people and the marine environment.[103]

## *Promoting Sustainable Aquaculture*

Celebrating recent developments like genetically altered fish, greater technological inputs, stronger antibiotics, and a faster, more reliable growing cycle, the World Bank in 1995 called aquaculture the "next great leap in food production." But caution is advised before jumping onto the fish-farm bandwagon. In the shift away from its traditional roots in Asia toward export-oriented commercial production, many farmers have lost sight of what makes aquaculture so appealing in the first place: its considerable potential for efficient use of resources and for integration with other methods of food production. Currently, some rapidly expanding practices are taking a toll on wild fish stocks, valuable coastal areas, and local people, undercutting the prospects of a sustainable industry. If aquaculture continues in the direction of highly intensive systems, fish farmers will risk running their own operations into the ground and further jeopardizing wild fisheries.[104]

Aquaculture has not always inflicted environmental harm. Indeed, integrated fish and rice farming has been the backbone of traditional agriculture in Asia for centuries. There, a system of raising several types of carp and other freshwater species together (polyculture) gradually evolved that drew on locally available resources. For food, the fish rely largely on the nutrient-rich residues and wastes from pigs or other animals, food preparation, nightsoil, or algae. In exchange, water in the fish pond provides backup for farm crops during the dry season. Plants grown in the fish ponds provide additional food for the fish and can also be harvested to feed pigs or other animals. Several times a year, the renewable rich bottom soil is scooped out of the fish ponds and applied to nearby fields, thus rounding out the on-farm organic loop.[105]

Traditional forms of fish farming, as practiced in Asian inland waters, offer enormous potential for local food security and household nutrition, a fact now being recognized by a growing number of development agencies, governments, and researchers. In recent projects, farmers in Vietnam report that raising fish in their rice paddies enables them to reduce fertilizer inputs by 28 percent. Similarly, farmers in the Philippines saved $25 per hectare on herbicides and pesticides by raising fish on their farms. Integrating fish and vegetable farming in the Mampong Valley, Ghana, helped farmers achieve impressive economic and ecological results, as measured by indicators of gross and net income, species diversity, number of recycling flows, and total farm production and efficiency. These farmers also doubled the protein in their families' diet to more than 120 percent of recommended levels.[106]

With a growth rate of 17 percent between 1993 and 1994 in low-income, food-deficit countries, and more than 10 percent worldwide, aquaculture constitutes one of the fastest-growing sectors in world food production. Production of fish and shellfish alone was valued at more than $36 billion in 1995, up from almost $10 billion in 1984. (In 1995, aquatic plants contributed an additional 7 million tons in output, worth nearly $6 billion.) Aquaculture is also making waves in meat markets: for every 5 kilograms of beef produced globally, there are now 2 kilograms of farm-raised fish. In the United States, farmed catfish now outranks veal, mutton, and lamb combined, a trend that is likely to be repeated soon for other species worldwide as aquaculture continues its steady climb in output.[107]

Globally, marine catches still account for 80 percent of fish production, but fish farmers are quickly altering the balance. For some species, farming already accounts for a large part of supplies. Today, for instance, nearly 40 percent of the salmon consumed have lived in captivity for more of their lives than they have lived in the wild, compared to 6 percent a decade ago. Likewise, 40 percent of the mollusks—which include oysters, clams, and mussels—and 65 percent of freshwater fish have lived mostly in farmed environments or in an aquaculture setting. Around the world, an estimated 85 percent of all farmed species are noncarnivorous, freshwater fish—such as tilapia, carp, and milkfish—and shellfish that eat low on the food chain.[108]

**Aquaculture constitutes one of the fastest-growing sectors in world food production.**

Raising fish in a controlled environment allows farmers to manipulate the growing cycle of fish and thereby increase yields. Research efforts aimed at breeding fish to resist disease, grow faster, taste better, and thrive in highly controlled environments have yielded impressive results. In Norway, researchers have raised salmon that, after only three generations, could tolerate being 10 times more crowded than their ancestors could. In 1993, researchers in the Philippines announced that they had successfully cultivated an "aquatic chicken," an inexpensive source of protein with minimal feed requirements that grows 60 percent faster than the wild native strains.[109]

The expansion of traditional pond systems that are well integrated with the local environment and within the bounds of available resources is being outpaced by the growth of very intensive monoculture systems raising predominantly carnivorous, highly profitable species that demand large amounts of feed, water, and fertilizers. Aquaculture is being transformed into a resource-intensive industry, rather than one that contributes to greater resource reuse and recycling. Many species are now raised for quick cash, with little thought given to where the inputs of water, feed, and land come from, where the fish go after leaving the farm, and what environmental costs are incurred in the process. Even China, which accounts for 62 percent of the world's total aquaculture production, is now pushing in the direction of more intense production and increased output.[110]

The most profitable commodity in aquaculture, and one of the most polluting, is shrimp. In the past decade, world production of giant tiger prawns jumped more than sixfold, to 3.5 million tons. In Bangladesh, Ecuador, Indonesia, and India, as well as the more established shrimp-farming countries of Thailand and China, shrimp culture comprised a $6.3 billion industry and yielded a major export product in 1995.[111]

Salmon farming is following a growth trajectory parallel to that of shrimp. Global production increased from 7,000 tons in 1980 to 500,000 tons worth more than $2 billion in 1995. Farming of salmon, a cold-blooded, temperate species, is concentrated primarily in northern countries, with Norway leading world production. Scotland, Canada, the United States, and, more recently, Chile round out the top five salmon-farming nations in the world. Together, shrimp and salmon farming accounted for just 6 percent of world aquaculture output in 1995, but 23 percent of its value.[112]

In the short term, intensive shrimp farming is highly profitable: in a year, an individual shrimp farmer can make up to $10,000 per hectare for intensive production rates of 4 to 5 tons per hectare. This compares to the roughly $1,000 per hectare that a species such as milkfish or carp generates. But these economic returns do not account for ecological—and economic—losses such as habitat degradation. By converting diverse ecosystems to simple ones, fish farmers and the public lose a host of ecological goods such as fish, shellfish, timber, charcoal, and other products. They also lose services that coastal ecosystems provide, such as filtering and purifying water, cycling nutrients, removing contaminants, and buffering the land from coastal storms and severe weather. A study of the Matang mangrove in Malaysia revealed that its value for coastal protection alone exceeded the value of farmed shrimp by 170 percent.[113]

Unfortunately, the areas best suited for cultivating fish often coincide with ones already used by wild fish stocks. Shrimp, for example, grow best in places with a mixed flow of fresh- and saltwater, such as the Gulf of Thailand and the Gulf of Fonseco in Honduras. Producers have tended to replicate these natural conditions by locating shrimp ponds in or near ecologically valuable coastal areas, where they can flood farmland with saltwater, making it useless for raising crops later on. Salmon cages have replaced shellfish beds in the low-lying shoreline areas in and around Puerto Montt, Chile, and antibiotics and high-nutrient feeds contaminate the water. Not all places where shrimp and fish ponds now appear were previously pristine—in some, shrimp cultivation succeeded agriculture, milkfish production, logging, or charcoal production. Nevertheless, fish farming has accelerated the pace of conversion and degradation of coastal habitats that are home to populations of wild shrimp, grouper, sea bass, crab, mullet, milkfish, and sea bream, among other species.[114]

Where fish are grown in artificial ponds, aquaculture can exacerbate pressures on available water resources, often competing with agriculture. Water serves as the medium for growth, replenishes oxygen, and removes wastes from the aquaculture system. In general, intensive production of common carp and tilapia requires roughly the same amount of water, per ton of fish, as the not-insignificant amounts needed to raise grain-fed cattle or pigs. (Raising 1 ton of fish consumes 8 tons of water, compared to 5 tons of water for a ton of pork and 8.5 tons for a ton of grain-fed beef.) But intensive shrimp production requires up to 10 times more water than most other fish species, creating enormous pressures on local water supplies.[115]

Displacing wild habitats has had measurable impacts on wild catches. In the coastal village of Ramachandrapuram, India, for instance, fishers reported a 10-fold drop in wild fish harvests one year after shrimp farms were built. And in one area of the Sundarbans, Bangladesh, fishers experienced an 80 percent decline in wild catches after mangroves were cut down and replaced with shrimp farms that altered fresh- and saltwater flows. The worst recorded damage has occurred in Thailand. Half of the mangrove forests in that country were lost during the 1980s, in large part due to shrimp-pond development. Although these ponds yielded 120,000 tons of farmed shrimp annually, they precipitated the loss of some 800,000 tons of wild fish harvests.[116]

When farmers cultivate species that eat other fish, such as eels, salmon, shrimp, or trout, they exacerbate pressures on wild stocks. Because carnivorous fish species require amino acids from other fish for growth, farmers feed them high-protein feed pellets made from wild fish. Cultivating salmon and shrimp requires 5 kilograms of wild ocean fish to be processed into fishmeal to produce 1 kilogram of farmed species. In inland Cambodia, Thailand, and southern Vietnam, fish farmers depend almost entirely on small, wild fish to feed their crops, species that would otherwise be dried or fermented and used as food by poor people. Consuming almost 20 percent of world fishmeal, aquaculture's high demand for this product is contributing to marine overfishing. Indeed, 6 million tons of wild ocean fish are used to feed farmed species each year. By 2010, it is expected that most of the world's fishmeal will be used to feed high-priced carnivorous fish on farms. But by and large, people would be better served if these fish went to feed people directly, not other fish.[117]

Aquaculture's threats to marine fisheries do not end with its demand for land and food inputs. Mismanaged fish farms can also "leak" into surrounding areas. Because fish grow in water, their waste and uneaten feed can float directly into the aquatic environment when no efforts are made to contain the refuse. Scottish researchers have estimated that each ton of cultivated fish can result in up to a ton of waste produced. Though generally not toxic, these nutrient-rich wastes can trigger eutrophication and the spread of algal blooms. Densely stocked salmon farms in British Columbia, Canada, were found to be producing an amount of waste equivalent to that generated by a city of half a million people.[118]

Another threat to wild species posed by aquaculture involves the escape of domesticated, farmed fish into rivers, lakes, and coastal areas. The fugitives can dilute the gene pool of wild species by eating the wild species or outcompeting them for food or displacing them altogether from particular areas. Norwegian scientists recently reported that one out of four salmon spawning in freshwater areas originally came from farms. Salmon are especially vulnerable to displacement, as they have territorial homing instincts that lead them back to site-specific rivers to spawn, a trait that is critical to their survival in the wild. The farmed salmon never develop this trait because they do not need it in cages. Once they interbreed with wild species, the trait is less common. The risks of displacement are even greater when the species under question is foreign to the area. Offshore storms in 1996 allowed nearly 100,000 farmed Atlantic salmon in Puget Sound to escape into the Pacific. These hardy non-native fish could undermine efforts by federal and state officials to restore endangered wild populations of Pacific salmon in the Columbia River and elsewhere.[119]

**People would be better served if fish went to feed people directly, not other fish.**

In addition to spilling over into the surrounding environment, poorly managed farms can also deplete the areas set aside for them. Over time, their considerable demands for feed (and often for electric power and other artificial inputs), combined with high levels of pollution and outbreaks of disease, sometimes lead to declining yields and, eventually, pond abandonment. Between 1965 and 1985, shrimp farmers in Taiwan intensified production per hectare from 1.4 tons of fish to 3.6 tons. But by the late 1980s, the aquaculture industry in Taiwan had completely collapsed due to a series of disease outbreaks and financial disasters, leaving behind the scars of a devastated landscape. More than 90 percent of shrimp ponds in the upper Gulf of Thailand, most of which are located on converted mangrove swamps, were deserted after two seasons because too much waste built up and clogged the pond. Between 1985 and 1995, approximately 150,000 hectares of shrimp ponds were abandoned worldwide, according to estimates by the World Wildlife Fund International. Even under the best circumstances, the average shrimp pond that receives large sums of food and fertilizer is only good for about 10 years.[120]

Governments are slowly waking up to the costs of destructive fish farming. To protect coastal areas, the government of Honduras implemented a one-year moratorium on new shrimp farms beginning in August 1996. The Supreme Court of India excluded all industrial shrimp farms from within 500 meters of the high-tide line that same year. Norway, too, has banned salmon netcages from fjords and coastal areas. In addition to limiting the worst aspects of the industry, governments are beginning to put more emphasis on its positive forms. The government of India, for example, subsidizes credit for small-scale aquaculture projects that provide food for local people.[121]

Given the rapid growth in aquaculture, a global strategy to encourage less-resource-intensive forms of production is needed. To date, production has been virtually unregulated and unchecked. Under the 1995 Code of Conduct for Responsible Fisheries, FAO analysts issued guidelines for sustainable aquaculture development. With support from national governments, scientists, environmentalists, and fish farmers themselves, these guidelines could help minimize aquaculture's impact on the environment. Signed by nearly 100 countries, a separate agreement on the role of fisheries in global food security—the 1996 Kyoto Declaration—calls for environmental and social impact statements for aquaculture operations, as well as for the use of native strains of fish.[122]

Clearly, the FAO guidelines and Kyoto Declaration do not obviate the need for more detailed and site-specific plans based on local conditions and concerns, but they do indicate a high level of concern, suggesting that many governments probably would support a global strategy on the issue. For the time being, zoning restrictions and minimum standards for resource integration and nutrient cycling can help minimize the drawbacks of aquaculture, while economic incentives and funding for clean production can help offset the runaway economics that currently favor the destructive practices of the aquatic gold rush.[123]

Because the sector is in flux, farmers, aquaculture officials, and others involved in raising fish have a critical opportunity to guide future developments and bring aquaculture back to its roots. But in order to restore health to the sector, at least three key changes are needed. First, it is important to encourage fish farmers to raise species—such as tilapia, carp, catfish, and shellfish—that require little or no fishmeal in their diet, in order to ease pressure on wild stocks. For farming, native strains of fish are preferable to exotic types—if they escape, they do not cause the usual problems of displacement.[124]

Second, halting the conversion of valuable coastal areas, especially ones that wild fish rely on, is imperative. Degrading these areas results in a loss of ecological goods and services, including valuable coastal protection services, and further undermines wild fish production. Finally, aquaculturalists need to come to terms with the pollution caused by their industry, to increase rates of resource recycling on fish farms, and to close the loop of fish production as best they can.[125]

As the world's expanding human population drives up global demand for new fish supplies and for protein sources generally, more people will look to aquaculture to provide a growing share of fish. Under favorable conditions, FAO estimates that aquaculture could sustainably supply up to 39 million tons of fish by 2010—about 70 percent more than is produced today. This would require increased investment, transfer of technology, and expansion of aquaculture outside traditional areas in Asia and Europe. But until the environmental problems plaguing the sector as a whole are addressed, as well as problems specific to the growth in species that eat other fish, farmers may have little success in realizing this potential.[126]

## *Fishing Laws Go Global*

Because fish swim across political boundaries and migrate without regard for management plans, regimes that oversee large marine areas or that seek to protect fish species throughout their entire range seem to offer the ideal unit from which to govern fisheries. Many such organizations currently exist and have met with varying degrees of success.

On the positive side, the South Pacific Forum Fisheries Agency, for example, banned large-scale driftnet fishing in the region in the late 1980s, thus helping to lay the groundwork for a global moratorium on this destructive gear in 1991. And the two-member International Pacific Halibut Commission, founded in 1923, has carried the species under its authority back from the brink of biological and economic disaster in the mid-1970s to a time of relatively high catches, reduced bycatch, and limited entry in the 1990s.[127]

Other authorities, however, such as the Pacific Salmon Commission, have failed to protect fisheries and have actually exacerbated conflict between different groups of fishers. And of course these problems become magnified when the species under question are highly migratory and valuable, such as tuna and swordfish, which cross between national EEZ boundaries and the open ocean. The International

Commission for the Conservation of Atlantic Tuna (ICCAT) has been called a "management nightmare" by one U.S. fisheries official. As a result of ongoing disagreements and lack of consensus within this 22-member commission, quotas have been set too high, leading to uncontrolled fishing by commercial and sportfishers and dramatic drop-offs in fish populations. Although conservation of highly migratory species is its raison d'être, ICCAT's main contribution to managing these fish has been in documenting their decline, according to Elliott Norse, president of the Marine Conservation Biology Institute.[128]

Because many populations of fish lack even minimal protection or oversight and some species span different regions, existing commissions and agreements have had mixed success in protecting fisheries. Truly effective fisheries conservation demands a global response. To date, however, world fisheries are covered by no comprehensive international regime or convention. Fortunately, a growing number of international agreements apply to particular aspects of the overfishing problem or to oceans management in general—most notably the 1982 U.N. Law of the Sea (UNCLOS). In response to rising concern for fisheries, a number of general international agreements and consensus statements slowly began to emerge in the early 1990s. UNCLOS entered into force on November 16, 1994, and within one year, a complementary agreement governing fisheries on the high seas was finalized by world fishing governments.[129]

Known as the "constitution of the oceans," UNCLOS established a comprehensive framework governing ocean use and setting such use in the context of environmental protection. In part, UNCLOS merely formalized what was already in practice and accepted as customary international law—most notably the right of national claims over the EEZ. But it also went far beyond existing practices. Within its 320 articles, nine annexes, and a 1994 supplementary agreement concerning seabed mining, the convention contains the building blocks for a precautionary approach to fisheries management. Specifically, these beginnings can be found in the requirements for environmental impact assessment and monitoring; caution in introducing technology and species; pollution control measures; and protection for rare, threatened, or endangered ecosystems and habitats.[130]

These conservation principles, however, are deeply embedded in the text and have proven difficult to apply, as the language and intent of the convention are often contradictory. Although the convention explicitly calls for protection of the marine environment within territorial waters (a coastal zone of up to 12 miles), for instance, obligations for fisheries conservation in this same zone remain unspecified. Related treaties have sought to clarify this and other ambiguities, but they have been oriented primarily towards development and use of fisheries, not their conservation and protection.[131]

One notable exception is the global moratorium on driftnet use. Prompted by the success of the regional ban in the South Pacific, the U. N. General Assembly passed a moratorium on high-seas driftnets in 1991. The decision to outlaw this indiscriminate fishing gear everywhere by June 1992 was prompted by public outrage over its environmental effects: extending up to 50 to 60 kilometers in the water, driftnets snare large quantities of marine mammals, seabirds, and fish. Outside of a few pockets of resistance, the use of gigantic driftnets today has virtually ended on the world's oceans. However, eliminating this particular type of gear has led fishers to use longlines and other damaging fishing methods to evade the specifics of the moratorium, often with similar effects on marine wildlife.[132]

Designed to complement the Law of the Sea, the oceans chapter of Agenda 21 (the plan of action endorsed by most of the world's governments at the 1992 Earth Summit) addresses the sustainable use and conservation of marine resources and habitat areas. Like the Law of the Sea, the language of this chapter with respect to fisheries conservation is weak and lacking in commitments for sustainable management. However, the global community will have an important opportunity to formalize tougher agreements when the U.N. Commission on Sustainable Development (the primary follow-up mechanism to the Earth Summit) addresses oceans and seas in 1999.[133]

Building on the general guidelines in UNCLOS and the oceans chapter of Agenda 21, more than 60 fishing countries agreed to a voluntary Code of Conduct for Responsible Fishing in 1995. This FAO-sponsored initiative sets forth principles for sustainable fisheries development, management, and conservation and calls for applying these principles to aquaculture as well. Bycatch and trade are also mentioned as key areas needing attention, but the code fails to note the role of subsidies. Many countries have endorsed the code, but evidence that it is making a difference—i.e., that more responsible fishing practices are taking hold—is scant.[134]

**Fishers now need to move rapidly into an era of precautionary management.**

The Convention on Highly Migratory and Straddling Stocks, which applies to all fish that straddle EEZ boundaries and migrate across the high seas, was also finalized in 1995. (Landings of migratory species—including the highly prized tuna, swordfish, and shark fisheries that have suffered from severe overexploitation—contributed nearly 16 million tons to world catches in 1994.) This convention marks the first international fisheries treaty or agreement to reject maximum sustainable yield as the standard for fisheries management, and the first to advocate a new standard: the precautionary principle. The precautionary principle holds that society should take action against certain practices when there is potential for irreversible consequences or for severe limits on the options for future generations—even when there is as yet no incontrovertible scientific proof that serious consequences will ensue. This precedent can help lead to more responsible fishing and set new standards for sustainable fishing. For most marine fisheries, however, severe damage has already occurred, and precautionary measures should be taken immediately. (See Table 3.)[135]

Under the terms of this convention, signatory states agree that their fleets will abide by the rules of applicable regional agreements on the high seas—even in cases where the signatory state is not party to that agreement. In the words of Satya Nandan, the chairman of the 1995 U.N. conference, "Governments will either cooperate to regulate fishing and conserve stocks, or they will not be allowed to fish." To ensure compliance, the convention creates the legal authority for state representatives of a different nationality from the country where the vessel is registered to board and inspect fishing vessels on the high seas.[136]

While establishing the precedent of the precautionary principle in international fisheries law, these tough provisions are not yet in force because the convention falls far short of the required 30 ratifications. Only four of the top 20 major fishing nations—Russia, Norway, Iceland, and the United States—have signed and ratified it. And six of the top 20 fishing states—Thailand, North Korea, Malaysia, Vietnam, Peru, and Chile—have not even signed the treaty, let alone ratified it.[137]

Because many of the major fishing nations of the world continue to evade responsibility for their actions at sea, fisheries conservation will benefit from a recent global agreement on the protection of marine biological diversity. The November 1995 meeting of the Conference of the Parties to the Convention on Biological Diversity (CBD) in Jakarta, Indonesia, adopted an action plan that calls for the protection of marine biological diversity and sustainable use of marine and coastal resources. The Jakarta Mandate, as it is known, puts the protection of marine fisheries in a broader context of biological and social goals. For instance, conserving salmon in the Pacific Northwest requires protecting entire ecosystems and sharing the benefits of salmon fishing equitably among fishers and local people. The work required to follow up and implement this action plan will not be easy. It will be necessary to define concepts clearly; to establish indicators for measuring progress; and to document suc-

**TABLE 3**

## Examples of Precautionary Measures

- Control access to the fishery early, before problems appear.
- Encourage responsible fishing through some form of fishing tenure or limited access.
- Place a cap on both fishing capacity and total fishing catch rate.
- Develop conservative catch limits and define upper range.
- If upper range is exceeded, implement recovery plans immediately to restore the stock.
- Reduce subsidies and encourage development of fisheries that are economically self-sufficient.
- Establish data collection and reporting systems.
- Avoid targeting fish that are too young or too small.
- Minimize bycatch through the use of more selective gear.
- Use area closures and marine protected areas to limit risks to the resource by providing refuges for stocks and restoring habitat.
- Develop management plans cooperatively with stakeholders and ensure ongoing participation and feedback.

*Source:* See endnote 135.

cessful case studies that can serve as useful models for national legislation and local work plans.[138]

Nevertheless, the Jakarta Mandate marks an important turning point in the way resource planners approach fisheries management—a shift that could help usher in a new body of law and agreements that grow primarily out of a concern for conservation rather than an exclusive focus on growth, development, and supply. If the Jakarta Mandate is effectively implemented, integrated coastal management, marine protected areas, and other conservation-oriented approaches will be more widely applied—to the benefit of global fisheries, coastal ecosystems, and marine biological diversity. Indeed, this is just what fish need. Without healthy habitats, their populations have little hope of recovering.[139]

While these recent policy initiatives help to fill the void in international law, a global consensus on the root problems of overcapacity and on subsidies has yet to emerge. More importantly, in nearly every agreement, the stated goal is conservation, but the intent and application have instead sanctioned business as usual: conservation and precautionary measures are left unspecified; specifics that are defined are negligible; enforcement is lax; and protection for species and marine habitat areas has generally been too little, too late to make a difference. Although these global commitments have been agreed to, many are still not ratified and thus lack the force of law. As the director general of the FAO, Jacques Diouf, warned in 1995, "the voluntary system of regulation of fisheries has not been successful."[140]

There is also a clear need for more fundamental change in the form of precautionary management. Current fisheries management regimes—which were designed to encourage development, growth, and maximum production—are inadequate to the task of addressing current challenges. Having witnessed the effects of overexploitation and fisheries stock collapses that accompany the overshooting of maximum sustainable yield, fisheries biologists, policymakers, and fishers now need to move rapidly into an era of precautionary management. In some cases, existing agreements have begun to build the foundation for this new era.[141]

## *Resetting the Compass*

Fish and fishers would benefit greatly if the urge to divide up the world's oceans and to control fisheries were tempered with ecological and social realities. Given the deteriorating condition of marine resources, fishers need to move quickly away from today's overcapitalized, heavily subsidized industry to a more sustainable, biologically sensitive, and socially diverse one. This transition will spark some short-term economic and social disruption, but will allow fisheries to recover and fishers to enjoy greater returns in the long run. A key issue is the extent to which society will guide these changes and help ease the short-term economic hardships—or merely let the chips fall where they may.[142]

One thing is clear: It is no longer enough for fishers simply to know how to operate a boat and catch fish. They need to be intimately involved in managing and conserving the resource. Whether the day-to-day tasks and responsibilities of fisheries management are devolved to the community level, shared between fishers and local or national governments (comanagement), relegated exclusively to private interests, or turned over to the state—or some combination of the above—bringing fishers together with scientists and resource planners is key to reconciling ecological realities with economic pressures to mine fisheries.[143]

Changes on two other broad fronts are also important prerequisites to sustainable fisheries management. First, it is critical to bring conservation back to the forefront of fisheries management. A more holistic, ecologically based approach that takes into account where fish live and where land-based activities take the greatest toll will go a long way toward renewing fishery resources. Second, economic measures and policy changes are necessary to reduce capacity and limit access. An obvious policy to achieve this aim is for governments to tackle the issue of reducing subsidies for fishing fleets and to overhaul economic incentives that bolster fishing capacity.[144]

To promote conservation, fishers and officials need to view fish as part of a larger ecological system, rather than simply as a commodity to extract. One tool that can help bring about this change is integrated coastal management (ICM). Through the process of community-based planning, ICM brings together diverse groups of people—fishers, politicians, tourism operators, traders, and the general public—to identify their shared problems and goals, and to define solutions that build on their common interests.

Discussions, mapping exercises, and site visits all help people make the connections between land and water use, and the health of fisheries.[145]

More than 10 years of experience with integrated coastal management in the Philippines has revealed the importance of approaching fisheries conservation as a series of steps in an ongoing process. To this end, it is important for fishers—and the community at large—to witness firsthand the connections between restored habitat and improved fish catches. Mangrove replanting projects and construction of artificial reefs are two concrete steps that help some fish stocks rebound quickly. Once people appreciate the immediate results of their work, they are more likely to engage in longer-term protection efforts, such as marine sanctuaries, which involve removing an area from use entirely.[146]

"No-take" fishing areas, seasonal fisheries closures, and marine protected areas also help depleted stocks rebound—and profits return—by limiting accessibility and easing pressures on the resource. (See Table 4.) A 1982 study of the offshore shrimp fishery in Texas waters documented a $9 million increase in the value of the shrimp catch following the closure of that fishery, which allowed juvenile shrimp to grow to more marketable sizes. In 1997, a group of international marine scientists called for governments to increase protected marine reserves from the current 0.25 percent of the ocean's surface area to 20 percent by 2020. Such a move would provide a much needed respite for commercially depleted fisheries and, combined with effective management of areas that continue to be fished, could allow global catches to grow by an estimated 10 million tons.[147]

In addition to habitat protection, fisheries conservation would benefit from policies and tools that help to restrict use of the resources—whether through marine protected areas or closed fishing seasons that apply to all fishers, or through more limited forms of entry such as Individual Transferable Quotas (ITQs). In an open access or shared resource, ownership does not materialize until fish are in the boat. ITQs, on the other hand, seek to extend ownership to fish while they are still in the water. However it is achieved, restricted use is no guarantee that fish will return. But if protective measures are imposed in an equitable and careful manner, people will more likely abide by the restrictions, thus dramatically improving the chances of recovery. Furthermore, reducing immediate pressures on the resource does buy time for fishers and policymakers to implement more drastic—and needed—cutbacks that will yield greater returns in the long run.[148]

Many governments are moving toward market-based systems for limiting access, of which ITQs are the most publicized. ITQs essentially privatize the right to fish by conferring on fishers a quota for a certain share of fish in a particular area and time which can be bought and sold on the market. Since individual shares rise or fall in proportion to changes in the total allowed catch, the theory is that fishers will go to greater lengths to protect the total stock, and thereby enjoy greater individual returns.[149]

Now implemented to varying degrees in Australia, Canada, Iceland, the Netherlands, New Zealand, and the United States, ITQs reduce overcapacity by enabling inefficient fishers to sell their quotas and thereby receive compensation for withdrawal from the industry. But this system of compensation has tended to concentrate quotas, and therefore access rights, in the hands of the few who can afford them. Just one year after the first ITQ program in the world was implemented in 1986 in New Zealand, the country's three largest fishing companies held title to 43 percent of the ITQs. Within five years, these three companies controlled half of all ITQs in the system. Many small-vessel owners and fishers either subcontracted to larger companies or went out of business completely.[150]

Modifications have since been applied in New Zealand, and more recent programs in other areas have attempted to address the social impacts of ITQs. Maori fishers in New Zealand now receive set shares of quotas, whereas Alaska has developed Community Development Quotas for communi-

**TABLE 4**

## Selected Marine Protected Areas and National Legislation

| Problem | Management Response | Results |
|---|---|---|
| **Apo Island Reef Sanctuary, Philippines** | | |
| Overfishing and destructive fishing methods damaging local fishing economy and coral reef ecosystem. Local fishers must travel 30 kilometers to find fish. | Marine Conservation and Development Program initiated in 1985 to provide environmental education, job training in other sectors, and water conservation measures. | By 1988, edible fish and shellfish populations recover. Fishers can fish in surrounding waters. |
| | Eight percent of reefs set aside in 1986 as a reserve limiting use to scuba diving and snorkeling. | After 10 years, reefs are intact, fish are bountiful, and rare species such as the valuable giant clam are thriving. |
| **Saba Island Marine Park, Lesser Antilles** | | |
| Mounting ecological impacts of the thriving tourist industry—including coral reef damage from boat anchors, sportfishing, diving—and from high extraction of marine organisms. Inadequate funding for enforcement and monitoring. | Locally designed, self-managed, and internally financed marine park established in 1987 to protect entire coastal zone. | Increase in visitors to marine park supports growth in related, locally based industries, including visitor services, buoy maintenance, and patrols. |
| | Reefs divided into four zones with a user fee for some zones: fishing and diving zone; anchoring zone; recreational diving zone; and multiuse zone for swimming, boating, snorkeling, diving, and fishing. | Divers charged user fee and tourists charged entry fee to park. |

| | | |
|---|---|---|
| Widespread illegal coral mining, degrading integrity of coral reef and contributing to severe beach erosion. | Coral mining and related activities banned; existing ban strengthened. | Forty-eight percent decrease in illegally mined coral between 1984 and 1993. |
| | Funding for alternative jobs for coral miners. | |
| | Public education for miners, police officers, and community to raise awareness of the damage from coral mining and to improve compliance with the law. | Special Area Management Plan enacted in Rekawa to incorporate sustainable use strategies for fisheries, agricultural land use, and beach habitat for sea turtles. |
| **Integrated Coastal Zone Management (ICZM) Program, Belize** | | |
| Overfishing causing declines in catches of several commercial species; frequent illegal fishing of undersized conch and lobster. | Institutional support for ICZM created through Technical and Steering Committees; monitoring, research, education, and training centers. No-fishing zone established. | Seven areas of protected reef throughout the country are UNESCO World Heritage Sites. |
| Uncontrolled tourism, including heavy boat traffic, sport fishing, and diving. | Fishermen, dive operators, and tour guides trained in conservation. | Belize's fishing and tourism industries becoming self-sufficient through tourism user fees. |
| Land-based activities causing severe run-off, pollution, and coastal erosion. | Legislative support embodied in the Coastal Zone Management Act. | Mooring buoys set to prevent reef damage; fishers monitor and assess stock. |

*Source:* See endnote 147.

ty-based, small-scale fishers. In Alaska, New Zealand, and other places, policymakers have set limits on consolidation of ownership to prevent monopolies from forming.[151]

The verdict is still out as to what extent ITQs confer conservation dividends. Although ITQs do tend to reduce pressure on individual fishing grounds and species, they impel fishers to move elsewhere. Quotas also encourage fishers to select only the highest-quality species, thereby increasing bycatch. Of the 31 ITQ fisheries reviewed by the Committee for Fisheries of the Organisation for Economic Co-operation and Development (OECD), most used other regulations, such as catch-size limits, gear restrictions, and closed seasons, to help mitigate the shortcomings of quota systems.[152]

A better way to protect fish may be to compensate governments directly for extraction of natural wealth. Charging fishers based on how much they take would also help pay for some of the costs of managing them. Many industrial countries now charge a flat fee for fishing permits and licenses; levying a tax on a particular gear type or practice would be a logical extension of this practice. Indeed, Australia now charges its domestic fishers about 2.5 percent of the value of landings earned at the dock. Worldwide, governments forgo between $3 and $7 billion in uncollected user fees from domestic and foreign fishers, according to estimates by Matteo Milazzo.[153]

In addition to charging fishers for the privilege of extracting the resource, reducing subsidies would also help protect fish stocks—and save governments and taxpayers money in the process. Economic programs that were originally intended to help fishers have helped destroy the resource by prolonging the notion that fishing is free. It is particularly important to engage the governments of China, the European Union, Japan, Norway, Russia, and the United States in a global subsidy reduction scheme because of their central role in providing subsidies. In contrast to the billions of dollars they spend propping up bloated, inefficient fleets, only about $500 million—5 percent of total subsidies from these governments—is budgeted to reduce fishing capacity.[154]

Of course, reducing overcapacity leads to questions about employment issues and what will happen to the people who rely on fishing for their jobs. In overcrowded industrialized fisheries, a combination of short-term economic aid to help excess laborers move out of the industry, and longer-term restructuring along with overall downsizing in the industry can help solve this problem.

In industrial countries, governments have paid to retrain fishers for other jobs or to buy back vessels, often with little success because they failed to implement the other crucial step: reducing the 95 percent of subsidies that motivate people to stay put and even enhance their capabilities. A nearly $2 billion five-year social adjustment program in Atlantic Canada is helping fishers stay out of debt for the time being, but critics argue that it amounts to nothing more than a massive social welfare program. In 1996, the European Union set aside $2.2 billion to pay for job-retraining programs and economic aid for out-of-work fishers. But intense pressure from industry officials, politicians, and fishers prompted EU ministers to backslide on fleet reductions and to postpone needed reforms.[155]

The challenge of finding work for displaced fishers is more difficult in developing countries because more people there rely on fishing for their livelihoods, alternative jobs are more limited than in industrial nations, and governments have scarce financial resources to devote to the problem. Also, in coastal communities from Atlantic Canada to Southeast Asia, fishing is not just a job, it is a way of life. In a survey of nine community-based coastal resource management projects in the Philippines, more than 80 percent of fishers said they would not leave the sector, even if they had a job with comparable income, because their lives and culture were so closely linked to the sea. Rather than trying to create jobs for people who want to stay put, some authorities are drawing on the tradition of part-time or seasonal involvement in fishing, with supplemental income from other activities. The result is that overall pressure on the fish is still reduced, but the wealth is spread over a larger number of people.[156]

Other communities hope to preserve fish as an important source of food and jobs by keeping large boats out of inshore and coastal areas where small-scale fishers tend to work. With nearly 30,000 islands covering an area greater than the continental United States, Indonesia has excluded medium- and large-scale fishers from an inshore zone since 1980. While intrusions into this zone do occur, the ban has protected an important way of life and has encouraged indigenous fishers to become more involved in patrolling their waters. Similarly, the more than 7 million small-scale fishers operating in India's coastal areas successfully pressured their government to prohibit joint ventures with foreign vessels in 1995, and the Supreme Court of Chile banned U.S. fishing trawlers from entering their waters in late 1997.[157]

Instead of regulating where people fish, some countries are now restricting how they fish. Namibia, for example, has enacted a no-discards law that prohibits throwing edible and marketable fish back to sea. The regulations apply across all sectors of the fishery—most of which operate under a quota system—and are enforced by onboard observers. Since implementation in the early 1990s, the policy has received widespread industry support and is viewed as highly effective at reducing discards. Norway passed similar legislation in 1996.[158]

U.S. fishery officials are now in the process of writing tougher bycatch laws that require the use of more selective gear. Turtle excluder devices (TEDs) costing $75–$500 apiece are required on trawl nets used to capture shrimp under the U.S. Endangered Species Act. Shaped like a funnel in the net that allows turtles to escape drowning, these devices have helped preserve the last Kemp's Ridley turtles in the Gulf of Mexico, as well as numerous finfish. Worldwide, investments in more selective gear types could yield reductions in discards of nearly 60 percent, resulting in an additional 15 million tons of fish landings per year, according to the FAO.[159]

Another issue that continues to undermine fisheries worldwide is the lack of enforcement and monitoring of existing agreements and bans. In the tropics, many communities are now working to educate fishers and the public about the dangers of cyanide fishing and to generate support for enforcing a ban on its use. For deep sea fisheries that are far removed from public oversight, Australia and New Zealand use satellite-based systems to monitor vessel movement and to make sure they do not move into no-fishing areas or grounds limited to national fishers. The SPFFA is planning to adopt a similar system to monitor the position of hundreds of vessels in South Pacific waters.[160]

**Worldwide, governments forgo between $3 and $7 billion in uncollected user fees.**

Ultimately, the success of all the management tools depends on accurate fish population data and conservative catch limits that take biological uncertainty into account. Movement in this direction, however, is hampered by the fact that most governments in developing countries lack a ministry of fisheries. Consequently, fishery issues come under the purview of the agriculture or commerce ministry, which does not consult the scientists and resource users who know the fish stocks and appreciate the inherent variability in a large-scale and complex ecosystem.[161]

One small but significant approach to addressing the need for better data comes from coral reef scientists and coastal managers, who have recently enlisted the help of recreational scuba divers. Numbering more than 7 million worldwide, sport divers who volunteer to collect data are given basic training to identify and survey fish and coral species, and to conduct rudimentary site assessments. The data are then compiled by scientists and put into a global inventory available on CD-ROM, known as ReefBase, which policymakers use to monitor the trends and conditions of reefs and to target intervention and protection programs. More efforts that engage the help of concerned individuals and volunteers would help overcome funding and data deficiencies and build greater public awareness of the problems plaguing world fisheries.[162]

In addition to on-site surveys, determining the effects of wide-scale changes in oceanographic conditions is a relatively new pursuit that deserves more attention. For example, more accurate models of ocean temperature, salinity, and currents are now being developed that can help officials better distinguish between changes due to overfishing and those due to natural stock migration or changing ocean and climate conditions. Officials can then adjust catch limits accordingly.[163]

Promoting sustainable fisheries also means addressing the growing demand for fish. To shift demand away from environmentally damaging fishing techniques and products, market forces, such as charging consumers a higher price for such selections, can be enlisted. A U.S. consumer boycott of canned tuna in the late 1980s successfully forced changes in the way fishers caught tuna and helped protect dolphins from being ensnared in purse seine nets. In April 1996, the World Wide Fund for Nature teamed up with one of the world's largest manufacturers of seafood products, Anglo-Dutch Unilever, to create economic incentives for sustainable fishing. Implemented through an independent Marine Stewardship Council, fisheries products that are harvested in a sustainable manner will qualify for an ecolabel. More efforts like these could help persuade manufacturers to replace fish oils and fishmeals with other products, to convince fishers to curb wasteful practices, and to generate greater public awareness of the need to carefully choose what fish species and products to eat.[164]

Consumer education can also tackle the growing demand for live fish and specialty products that result in high environmental costs. People who enjoy shark fin soup, live fish prepared before their eyes, and other aquatic delicacies can be charged the full ecological price for their meals, for it is their demand that drives the use of deadly practices such as cyanide fishing.[165]

Ultimately, what determines the shape that fishing will take in the future begins at home—far from the high seas and underwater habitats that are under siege. Consumers can have a positive impact on global fisheries by buying fish products that have been made sustainably, asking where a fish came from and how it was captured or raised, and demanding that policymakers support the recommendations of scientists to close fisheries and reduce the amount of fishing.[166]

The current habit of reacting to fishery crises rather than preventing biological declines cannot continue indefinitely. With fewer and fewer commercially viable species available, fishers will soon face catastrophic losses from which few governments will be able to rescue them. By shifting our focus from what is done to fish to what can be done for fish, fisheries can continue to provide food, jobs, and enjoyment for hundreds of millions of people worldwide. Ultimately, such a shift can bring world fisheries back within the bounds of nature and allow this renewable resource to flourish.

# Notes

1. Anthony DePalma, "Canadians Block U.S. Ferry in a Salmon-Fishing Fight," *New York Times*, 22 July 1997; "Fishermen Release Alaskan Ferry," *Providence Journal-Bulletin*, 23 July 1997.

2. Ross Howard and Craig McInnes, "Alaskan Catch Angers Clark," *Globe and Mail* (Toronto), 18 July 1997; Carey Goldberg, "Fishing Talks and Fighting Words," *New York Times*, 5 June 1997.

3. Forty percent and 300 salmon from Carl Safina, *Song for a Blue Ocean: Encounters Along the World's Coasts and Beneath the Seas* (New York: Henry Holt and Company, 1997).

4. Farmed salmon and the effects on prices from Howard Schneider, "The Fish that Spawned a Crisis Is Not Exactly on Apec's Menu," *Washington Post*, 25 November 1997; Statue of Liberty quote from Timothy Egan, "Salmon War in Northwest Spurs Wish for Good Fences," *New York Times*, 12 September 1997.

5. Scott Morrison, "US Sued in Salmon War," *Financial Times*, 10 September 1997.

6. U.N. estimate from Jonathan Friedland, "Fish Stories These Days Are Tales of Depletion and Growing Rivalry," *Wall Street Journal*, 25 November 1997.

7. Note: The word fishery has several different meanings: it can refer to a particular fishing area, such as the South China Sea; to a group of fish species, such as cod or orange roughy; or, more broadly, to the area where fishing occurs and the mix of species captured. In this paper, the latter definition is meant unless otherwise specified. Marine fish stocks and fishing grounds from U.N. Food and Agriculture Organization (FAO), *The State of World Fisheries and Aquaculture, 1996* (Rome: 1997); lower quality and value from Daniel Pauly et al., "Fishing Down Marine Food Webs," *Science*, 6 February 1998; one fourth from S. M. Garcia and C. Newton, "Current Situation, Trends, and Prospects in World Fisheries," in E.K. Pikitch, D.D. Huppert, and M.P. Sissenwine, *Global Trends: Fisheries Management*, American Fisheries Society (AFS) Symposium 20 (Bethesda, MD: AFS, 1997); 1995 aquaculture estimates from FAO, *Aquaculture Production Statistics, 1986–1995*, FAO Fisheries Circular No. 815, Revision 9 (Rome: 1997); 1984 estimate from FAO, *Aquaculture Production Statistics, 1984–1993*, FAO Fisheries Circular No. 815, Revision 7 (Rome: 1995).

8. Twice capacity from Chris Newton and John Fitzpatrick, *Assessment of the World's Fishing Fleet, 1991–1997* (Washington, DC: Greenpeace, 1998); one third revenues comes from Matteo Milazzo, *Subsidies in World Fisheries: A Reexamination*, World Bank Technical Paper No. 406, Fisheries Series

(Washington, DC: World Bank, April 1998).

**9.** Areas that serve as natural fish farms from Don Hinrichsen, "Coral Reefs in Crisis," *BioScience*, October 1997, quoting Rodney Salm, Marine and Coastal Conservation Coordinator, World Conservation Union (IUCN), Nairobi, Kenya; Tundi Agardy, "It's Not Just What We Fish, But How We Fish that Matters," *Christian Science Monitor*, 6 March 1998.

**10.** Trade from FAO, *The State of World Fisheries and Aquaculture, 1996*, op. cit. note 7; George Kent, "Fisheries, Food Security, and the Poor," *Food Policy*, vol. 22, no. 5, 1997; 1 billion people from "Water and Fisheries," in World Resources Institute (WRI) et al., *World Resources 1996–97* (Washington, DC: Oxford University Press, 1996).

**11.** Bob Holmes, "Blue Revolutionaries," *New Scientist*, 7 December 1996; "slash and burn" analogy from "Trimming the Fleet: Can Wild Fisheries Recover?" *Coastal Heritage*, fall 1996.

**12.** History of freedom of the seas from John Cordell, ed., *A Sea of Small Boats* (Cambridge, MA: Cultural Survival, Inc., 1989).

**13.** Thousands of years from Jeffrey S. Walters, "Coastal Common Property Regimes in Southeast Asia," in Elisabeth Mann Borgese, Norton Ginsburg, and Joseph R. Morgan, eds., *Ocean Yearbook*, vol. 11 (Chicago: University of Chicago Press, 1994); John Cordell, "Introduction to Sea Tenure," in Cordell, op. cit. note 12; James R. McGoodwin, *Crisis in the World's Fisheries: People, Problems, and Policies* (Stanford, CA: Stanford University Press, 1990); broad belief system from Moana Jackson, "Indigenous Law and the Sea," in Jon M. Van Dyke, Durwood Zaelke, and Grant Hewison, eds., *Freedom for the Seas in the 21st Century: Ocean Governance and Environmental Harmony* (Washington, DC: Island Press, 1993).

**14.** "High degree of power sharing" from Evelyn Pinkerton and Martin Weinstein, *Fisheries That Work: Sustainability through Community-Based Management* (Vancouver, BC: David Suzuki Foundation, July 1995). See also Elinor Ostrom, *Governing the Commons: The Evolution of Institutions for Collective Action* (New York: Cambridge University Press, 1990); T.R. McClanahan, H. Glaesel, J. Rubens, and R. Kiambo, "The Effects of Traditional Fisheries Management on Fisheries Yields and the Coral-Reef Ecosystems of Southern Kenya," *Environmental Conservation*, no. 2, 1997.

**15.** W.L. Hobart, ed., *Baird's Legacy: The History and Accomplishments of NOAA's National Marine Fisheries Service, 1871-1996*, National Oceanic and Atmospheric Administration (NOAA) Technical Memorandum NMFS-F/SPO-18 (Silver Spring, MD: NOAA, National Marine Fisheries Service (NMFS), December 1995).

**16.** New England today from Conservation Law Foundation, "13 New

England Stocks Declared Overfished," *New England Fisheries* (Boston, MA: October 1997); see also Okon Akiba, "Policy Issues and Challenges in Canadian Management of the Atlantic Fisheries," *Environmental Conservation*, no. 2, 1997; growth in marine catches from R.J.R. Grainger and S.M. Garcia, *Chronicles of Marine Fishery Landings (1950–1994): Trend Analysis and Fisheries Potential*, FAO Fisheries Technical Paper No. 359 (Rome: FAO, 1996).

**17.** World's major fishing areas excludes inland catches and Antarctic waters. Eleven of 15 Worldwatch estimate based on data from Maurizio Perotti, fishery statistician, Fishery Information, Data and Statistics Unit (FIDI), Fisheries Department, FAO, Rome, e-mail message to author, 14 October 1997; 60 percent from FAO, op. cit. note 7; reduce capacity and rehabilitate from Grainger and Garcia, op. cit. note 16.

**18.** Figure 1: 1997 preliminary data from Perotti, FIDI, letter to author, 23 March 1998; data for 1984–96 from Perotti, FIDI, letter to author, 22 March 1998; 1950–84 world production and total catch from FAO, *Yearbook of Fishery Statistics: Catches and Landings* (Rome: 1967–91); 1984–90 culture data from FAO, *Aquaculture Production Statistics, 1984–93*, op. cit. note 7; before 1984, culture estimates are based on a 1975 world aquaculture production estimate from National Research Council, *Aquaculture in the United States: Constraints and Opportunities* (Washington, DC: National Academy of Sciences, 1978) and on country estimates in Conner Bailey and Mike Skladeny, "Aquacultural Development in Tropical Asia," *Natural Resources Forum*, February 1991; growth rates from FAO, *Marine Fisheries and the Law of the Sea: A Decade of Change*, FAO Fisheries Circular No. 853 (Rome: 1993); one out of five fish today from Holmes, op. cit. note 11; one out of four from Manny Ratafia, "Aquaculture Today: A Worldwide Status Report," *World Aquaculture*, June 1995; overtook beef and poultry from Lester R. Brown et al., *Vital Signs 1998* (New York: W. W Norton & Company, 1998).

**19.** Simon Fairlie, Mike Hagler, and Brian O'Riordan, "The Politics of Overfishing," in "Overfishing: Causes and Consequences," *The Ecologist*, March/April, May/June 1995; growth rate from Grainger and Garcia, op. cit. note 16; for further discussion see Peter Weber, *Net Loss: Fish, Jobs and the Marine Environment*, Worldwatch Paper No. 120 (Washington, DC: July 1994).

**20.** Lisa Speer et al., *Hook, Line and Sinking: The Crisis in Marine Fisheries* (New York: Natural Resources Defense Council, February 1997); sharks from Marjorie L. Mooney-Seus and Gregory S. Stone, *The Forgotten Giants: Giant Ocean Fishes of the Atlantic and Pacific* (Washington, DC: Ocean Wildlife Campaign, 1997); decline in Japanese catches from FAO, *Yearbook of Fishery Statistics: Catches and Landings*, vol. 80 (Rome: 1997), and from "News Round-Up: Dwindling," *Samudra*, July 1996; fishing area declines from Perotti, op. cit. note 17.

**21.** Mackerel and herring from "Introduction," *The Ecologist*, March/April,

May/June 1995; herring also from David J. Whitmarsh et al., "Natural Resources Exploitation and the Role of New Technology: A Case-History of the UK Herring Industry," *Environmental Conservation*, summer 1995; R.M. Cook, A. Sinclair, and G. Stefansson, "Potential Collapse of North Sea Cod Stocks," *Nature*, 6 February 1997; "Disaster Is Feared in Fish Population," *International Herald Tribune*, 15–16 March 1997.

**22.** Gulf of Thailand from Boonlert Phasuk, "Fishing Effort Regulations in the Coastal Fisheries of Thailand," in Indo-Pacific Fishery Commission (IPFC), *Socio-Economic Issues in Coastal Fisheries Management*, Proceedings of the IPFC Symposium (Bangkok: FAO, 1994); Bolinao sea urchin example from L. Talau-McManus and K.P.N. Kesner, "Valuation of a Philippines Municipal Sea Urchin Fishery and Implications of Its Collapse," in Marie Antoinette Juinio-Meñez and Gary F. Newkirk, eds., *Philippine Coastal Resources Under Stress: Selected Papers from the Fourth Annual Common Property Conference* held in Manila, Philippines, June 16–29, 1993 (Manila: International Center for Living Aquatic Resources Management (ICLARM), 1995).

**23.** David Malakoff, "Extinction on the High Seas," *Science*, 25 July 1997; Cheryl Lyn Dybas, "A Tale of Two Fishes," *Wildlife Conservation*, March/April 1998.

**24.** General trend from Pauly et al., op. cit. note 7; New England and mid-Atlantic U.S. region examples from Speer et al., op. cit. note 20; Figure 2 from NMFS, *Our Living Oceans: The Economic Status of U.S. Fisheries* (Silver Spring, MD: December 1996).

**25.** Pauly et al., op. cit. note 7.

**26.** Five pelagic species from FAO, *Review of the State of World Fishery Resources, Part 1: Marine Resources*, FAO Fisheries Circular No. 710, Revision 8, Part 1 (Rome: 1992); figure of 73 percent is Worldwatch estimate based on data in FAO, *Marine Fisheries and the Law of the Sea*, op. cit. note 18; five pelagic and four demersal species from Garcia and Newton, op. cit. note 7.

**27.** Swordfish data from Speer et al., op. cit. note 20; quote from William K. Stevens, "Long-Line Fishing Seen as Damaging to Some Fish and to the Albatross," *New York Times*, 5 November 1996. For more information on damages from longlines and other gear, see Ocean Wildlife Campaign, *Indiscriminate Slaughter at Sea: A Primer on Bycatch* (Washington, DC: January 1998).

**28.** Caribbean from Callum M. Roberts, "Effects of Fishing on the Ecosystem Structure of Coral Reefs," *Conservation Biology*, October 1995; Julie P. Hawkins and Callum M. Roberts, "The Growth of Coastal Tourism in the Red Sea: Present and Future Effects on Coral Reefs," *Ambio*, December 1994; Kenya from J.W. McManus, "Tropical Marine Fisheries and the Future of Coral Reefs: A Brief Review with Emphasis on Southeast Asia," *Coral*

*Reefs*, vol. 16, no. 5, 1997.

**29.** "The Black Sea in Crisis," *Environmental Health Perspectives* (Research Triangle, NC: National Institute of Environmental Health Science, December 1997); *Mnemiopsis ledyi* from Elliott Norse, President, Marine Conservation Biology Institute (MCBI), Redmond, Washington, e-mail message to author, 22 April 1998.

**30.** Forty percent and particular species from Grainger and Garcia, op. cit. note 16; impossible to catch all species at peak levels at the same time from Garcia and Newton, op. cit. note 7.

**31.** Examples of species that have recovered from Pamela M. Mace, "Developing and Sustaining World Fisheries Resources: The State of the Science and Management," Keynote Address, Second World Fisheries Congress, Brisbane, Australia, 28 July 1996, in D.A. Hancock, D.C. Smith, A. Grant, and J.P. Beumer, eds., *Developing and Sustaining World Fisheries Resources: The State of the Science and Management* (Victoria, Australia: CSIRO Publishing, April 1997); Daniel Pauly and Villy Christensen, "Rehabilitating Fished Ecosystems: Insights from the Past," *NAGA, The ICLARM Quarterly*, July 1996.

**32.** Resource exhaustion from Ferren MacIntyre, Kenneth W. Estep, and Thomas T. Noji, "Is It Deforestation or Desertification When We Do It to the Ocean?" *NAGA, The ICLARM Quarterly*, July 1995; Hobart, op. cit. note 15.

**33.** Scott Sonner, "Scientists: Sorry State of World's Oceans Are a Warning to Humans," *Corvallis Gazette-Times*, 25 January 1998; Associated Press, "1,600 Scientists Call for Action to Save the Oceans," *Providence Journal-Bulletin*, 7 January 1998; MCBI, "Troubled Waters: A Call to Action," <http://www.mcbi.org>, viewed 7 May 1998; Quote from "Statement of Dr. M. Patricia Morse, Professor of Biology, Northeastern University, on Release of Troubled Waters: A Call to Action," <http://www.mcbi.org>, viewed 7 May 1998.

**34.** Rates of marine biological productivity from different areas of the ocean from D. Pauly and V. Christensen, "Primary Production Required to Sustain Global Fisheries," *Nature*, 16 March 1995; 80–90 percent from Cordell, op. cit. note 12.

**35.** For discussion of importance of coastal habitat to fish, see Boyce Thorne-Miller and John Catena, *The Living Ocean: Understanding and Protecting Marine Biodiversity* (Washington, DC: Island Press, 1991); Bay of Bengal, East Africa, and Sri Lanka from Martin W. Holdgate, "The Sustainable Use of Tropical Coastal Resources—A Key Conservation Issue," *Ambio*, November 1993; menhaden from Susan-Marie Stedman and Jeanne Hanson, "Wetlands, Fisheries, and Economics in the Gulf of Mexico Coastal States," *Habitat Connections*, vol. 1, no. 4, 1997.

**36.** Asia from Don Anderson, "Turning Back the Harmful Red Tide," *Nature,* 7 August 1997; Baltic problems from Jonas Gunnarsson et al., "Interactions Between Eutrophication and Contaminants: Towards a New Research Concept for the European Aquatic Environment," *Ambio,* September 1995; toxic and nontoxic blooms from Donald B. Anderson, "Red Tides," *Scientific American,* August 1994; see also William J. Broad, "A Spate of Red Tides Menaces Coastal Seas," *Washington Post,* 27 August 1996; Mississippi River from Mark Schleifstein, "Fertilizer, Sewage Brew Dead Zone," *New Orleans Times-Picayune,* 25 March 1996; to size reference from Joby Warrick, "'Dead Zone' Plagues Gulf Fishermen: Stagnant Waters Reach 6,800 Square Miles," *Washington Post,* 24 August 1997; Puerto Rico from *The New York Times Atlas of the World,* Third Revised Concise Edition (New York: Bartholomew/Times, 1992).

**37.** Share of population near coastline from Joel E. Cohen et al., "Estimates of Coastal Populations," *Science,* 14 November 1997; modified original estimates of 60 percent from Peter M. Vitousek et al., "Human Domination of Earth's Ecosystems," *Science,* 25 July 1997; 3.8 billion from Don Hinrichsen, "Humanity and the World's Coasts: a Status Report," *Amicus Journal,* winter 1997; coastal cities and 6.3 billion from Don Hinrichsen, "Coasts in Crisis," *Issues in Science and Technology,* summer 1996; Michael H. Glantz, *The Impacts of Climate on Fisheries,* United Nations Environment Program (UNEP), Environment Library No. 13 (Nairobi: UNEP, 1994).

**38.** Janet Raloff, "Fishing for Answers: Deep Trawls Leave Destruction in Their Wake—But for How Long?" *Science News,* 26 October 1996; Dick Russell, "Hitting Bottom," *Amicus Journal,* winter 1997; North Sea and France from Robert T. Paine, "Coasts in Crisis," *Issues in Science and Technology,* fall 1996; Georges Bank from Peter J. Auster et al., "The Impacts of Mobile Fishing Gear on Seafloor Habitats in the Gulf of Maine (Northwest Atlantic): Implications for Conservation of Fish Populations," *Reviews in Fisheries Science,* no. 2, 1996; "parking lot" comment from Jeremy Collie, University of Rhode Island, Graduate School of Oceanography, Narragansett, RI, presentation at 1997 Annual Meeting of the Society for Conservation Biology, University of Victoria, British Columbia, 7 June 1997; Elliott A. Norse, "Bottom-Trawling: The Unseen Worldwide Plowing of the Seabed," *The NEB Transcript* (Beverly, MA: New England Biolabs, Inc., January 1997).

**39.** Charles Victor Barber and Vaughan R. Pratt, *Sullied Seas: Strategies for Combating Cyanide Fishing in Southeast Asia and Beyond* (Washington, DC: WRI and International Marinelife Alliance, August 1997); Mark V. Erdmann and Lida Pet-Soede, "How Fresh Is Too Fresh?: The Live Reef Food Fish Trade in Eastern Indonesia," *NAGA, The ICLARM Quarterly,* January 1996.

**40.** Discards of 20 million tons from David Wilmot, Director, Ocean Wildlife Campaign, Washington, DC, e-mail message to author, 12 February 1998, and American Sportfishing Association and Ocean Wildlife

Campaign, "Slaughter at Sea," press release (Washington, DC: 12 January 1998); global bycatch from Dayton L. Alverson et al., *A Global Assessment of Fisheries Bycatch and Discards*, FAO Fisheries Technical Paper 339 (Rome: FAO, 1994).

**41.** Mike Hagler, "Deforestation of the Deep," *The Ecologist*, March/April, May/June 1995; John H. Cushman Jr., "Cuts Sought in Wasteful Fish Kills," *New York Times*, 13 January 1998.

**42.** Cost data from Alverson et al., op. cit. note 40; data for 1995 Northwest Pacific catches from FAO, op. cit. note 20.

**43.** Carl Safina, "The World's Imperiled Fish," *Scientific American*, November 1995; Brazil and worldwide from Alverson et al., op. cit. note 40.

**44.** Louis W. Botsford, Juan Carlos Castilla, and Charles H. Peterson, "The Management of Fisheries and Marine Ecosystems," *Science*, 25 July 1997; fish cannot regulate their internal temperature and climate effects from Robert T. Watson et al., eds., *Climate Change 1995—Impacts, Adaptations, and Mitigation of Climate Change: Scientific-Technical Analyses*, Contribution of Working Group II to the Second Assessment Report of the Intergovernmental Panel on Climate Change (IPCC) (Cambridge, U.K.: Cambridge University Press, 1995).

**45.** Lewis M. Rothstein and Dake Chen, "The El Niño/Southern Oscillation Phenomenon," *Oceanus*, fall/winter 1996; 1997 effects from Associated Press, "El Niño Heats Waters to Record Highs in West," *New York Times*, 4 September 1997; Bering Sea from Chris Bright, "Tracking the Ecology of Climate Change," in Lester R. Brown, Christopher Flavin, and Hilary French, *State of the World 1997* (New York: W.W. Norton & Company, 1997).

**46.** John McGowan and Dean Roemmich, "Climatic Warming and the Decline of Zooplankton in the California Current," *Science*, 3 March 1995; Richard R. Veit et al., "Apex Marine Predator Declines Ninety Percent in Association with Changing Oceanic Climate," *Global Climate Biology*, vol. 3, 1997.

**47.** IPCC predictions from Watson, op. cit. note 44; David Schneider, "The Rising Seas," *Scientific American*, March 1997; Glantz, op. cit. note 37; shiner from Bright, op. cit. note 45.

**48.** "Fish Damage Linked to UV," *New York Times*, 18 March 1997; Donat-P. Häder, Robert C. Worrest, H.D. Kumar, and Raymond C. Smith, "Effects of Increased Solar Ultraviolet Radiation on Aquatic Ecosystems," *Ambio*, 3 May 1995.

**49.** Forty million tons from Garcia and Grainger, op. cit. note 16.

**50.** Twice capacity from Newton and Fitzpatrick, op. cit. note 8; Weber, op. cit. note 19.

**51.** Karyn L. Gimbel, ed., *Limiting Access to Marine Fisheries: Keeping the Focus on Conservation* (Washington, DC: Center for Marine Conservation and World Wildlife Fund (WWF), 1994); NMFS, op. cit. note 24.

**52.** Not inevitable from Pinkerton and Weinstein, op. cit. note 14.

**53.** Rögnvaldur Hannesson, *Fisheries Mismanagement: The Case of the North Atlantic Cod* (Cambridge, MA: Fishing News Books, Blackwell Science, Inc., 1996); costs of fishing from Pinkteron and Weinstein, op. cit. note 14; costs also from Brian J. Rothschild, "How Bountiful Are Ocean Fisheries?" *Consequences*, no. 1, 1996.

**54.** Garrett Hardin, "The Tragedy of the Commons," *Science*, 13 December 1968; unmanaged conditions from Garrett Hardin, "Extension of the 'Tragedy of the Commons'," *Science*, 1 May 1998; see also Pinkerton and Weinstein, op. cit. note 14.

**55.** Lennox Hinds, "World Marine Fisheries: Management and Development Problems," *Marine Policy*, September 1992. For discussion of bioeconomic model of fish populations see NMFS, op. cit. note 24.

**56.** David W. Pearce and Jeremy J. Warford, "Property Rights Failure and Renewable Resources," in *World Without End: Economics, Environment, and Sustainable Development* (New York: Oxford University Press, 1993); setting MSY from Donald Ludwig, Ray Hilborn, Carl Walters, "Uncertainty, Resource Exploitation, and Conservation: Lessons from History," *Science*, 2 April 1993, and A.A. Rosenberg et al., "Achieving Sustainable Use of Renewable Resources," *Science*, 5 November 1993; Thorolfur Matthiasson, *Why Fishing Fleets Tend to Be "Too Big,"* Iceland Economic Papers No. 38 (Réykjavik: Faculty of Economics and Business Administration, University of Iceland, January 1996); Hannesson, op. cit. note 53; today's situation from Rory McLeod, *Seafood Trade Access Study* (Wellington: New Zealand Fishing Industry Board, 1996).

**57.** Michael L. Weber and Judith A. Gradwohl, *The Wealth of Oceans* (New York: W.W. Norton & Company, 1995); see also McGoodwin, op. cit. note 13.

**58.** Weber and Gradwohl, op. cit. note 57; Ken Stump and Dave Batker, *Sinking Fast: How Factory Trawlers Are Destroying U.S. Fisheries and Marine Ecosystems* (Washington, DC: Greenpeace, August 1996).

**59.** Stump and Batker, op. cit. note 58.

**60.** Weber and Gradwohl, op. cit. note 57. For situation today in the United States see Gerry Leape, "Big Fishing Boats Threaten Fish Stocks and Jobs," press release (Washington, DC: Greenpeace, 5 May 1998).

**61.** R.P. Anand, "Changing Concepts of Freedom of the Seas: A Historical Perspective," in Van Dyke, Zaelke, and Hewison, op. cit. note 13.

**62.** Ibid.; UNCLOS from Terry D. Garcia and Jessica A. Wasserman, "The U.N. Convention on the Law of the Sea: Protection and Preservation of the Marine Environment," *Renewable Resources Journal*, spring 1995; David M. Dzidzornu, "Coastal State Obligations and Powers Respecting EEZ Environmental Protection Under Part XII of the UNCLOS: A Descriptive Analysis," *Colorado Journal of International Environmental Law and Policy*, no. 2, 1997.

**63.** Claudia Carr, "The Legacy and Challenge of International Aid in Marine Resource Development," in Van Dyke, Zaelke, and Hewison, op. cit. note 13; Asian Development Bank (ADB), *The Bank's Policy on Fisheries* (Manila: October 1997); U.S. Agency for International Development, *The Status of Fisheries and Aquaculture Development Assistance Programs* (Washington, DC: August 1993); Peter O'Neill, "Farming Has Expanded on World Bank Millions," *Fish Farming International*, July 1994.

**64.** Fleet growth in developing countries from Garcia and Newton, op. cit. note 7; Asia's share from Trish Saywell, "Fishing for Trouble," *Far Eastern Economic Review*, 13 March 1997; modern deep sea fleets from ADB, op. cit. note 63.

**65.** Sebastian Mathew, "Coastal Communities and Fishworkers: Factors in Fisheries Laws and Management," presentation at the Law of the Sea Institute, 30th Annual Conference, Al-Ain, United Arab Emirates, 20 May 1996, citing C. Sanger, *Ordering the Oceans: The Making of the Law of the Sea* (London: Zed Books, 1986); producers in 1994 from FAO, op. cit. note 20.

**66.** Figure 3 from Garcia and Newton, op. cit. note 7; all data are for vessels longer than 24 meters and larger than 100 gross registered tons. These vessels account for an estimated 1.7 percent of the total number of decked vessels, but almost 60 percent of total vessel tonnage in world fleets. For further discussion see FAO, *Marine Fisheries and the Law of the Sea*, op. cit. note 18, and Newton and Fitzpatrick, op. cit. note 8.

**67.** Garcia and Newton, op. cit. note 7; late 1980s from FAO, *State of World Fisheries and Aquaculture*, op. cit. note 7.

**68.** Situation since 1989 from Garcia and Newton, op. cit. note 7; 1992–97 data from Newton and Fitzpatrick, op. cit. note 8; other fisheries from Mace, op. cit. note 31.

**69.** Garcia and Newton, op. cit. note 7.

**70.** FAO, *Marine Fisheries and the Law of the Sea*, op. cit. note 18; pressure on governments from Norman Myers, *Perverse Subsidies: Tax $s Undercutting Our Economies and Environments Alike* (Winnipeg: International Institute for

Sustainable Development, 1998.

**71.** Worldwatch estimates based on FAO, *Marine Fisheries and the Law of the Sea*, op. cit. note 18.

**72.** Subsidies from Milazzo, op. cit. note 8; quote from Matteo Milazzo, Fisheries Economist, Office of Sustainable Fisheries, NMFS, NOAA, Silver Spring, MD, discussion with author, 18 March 1998; more likely estimate of global subsidies is between $20 and $30 billion, representing 25 and 37.5 percent respectively of global revenues ($80 billion in 1995) as estimated by Milazzo.

**73.** Gareth Porter, "Fisheries Subsidies, Overfishing and Trade," draft paper for UNEP in *Environment and Trade Series*, May 1996. For broader discussion of environmental subsidies see David Malin Roodman, *Paying the Piper: Subsidies, Politics, and the Environment*, Worldwatch Paper No. 133 (Washington, DC: Worldwatch Institute, December 1996).

**74.** U.S. estimates from Michael P. Sissenwine and Steven L. Swartz, "Shifting Currents in Fisheries Policy," *Forum for Applied Research and Public Policy*, summer 1996, citing data from NOAA 1995–2005: Strategic Plan (Washington, DC: NOAA, 1993); $15–30 billion from Deborah Cramer, "Troubled Waters," *Atlantic Monthly*, June 1995; up to 40 million tons from Grainger and Garcia, op. cit. note 16. Note: this estimate includes an estimated 6 million tons in cultured species from marine waters (mariculture), full development of unutilized stocks, and better management of stocks. The authors, Grainger and Garcia, caution that the estimate is highly unreliable, and increases of 20 million tons worldwide are more likely, given current levels of exploitation.

**75.** Coalition for Fair Fisheries Agreements (CFFA), *Fishing for a Future: Artisanal Fishing in Senegal and EU Fisheries Agreements* (Brussels: CFFA and CREDETIP, October 1995); Stephen Buckley, "Senegal's Fishermen Struggle to Pull a Living from the Sea," *Washington Post*, 1 November 1997; Michael Belliveau, "The Enemy Is Not Just 'Out There'," *Samudra*, July 1997; population of Senegal in 1997 from Don Hinrichsen, *Coastal Waters of the World: Trends, Threats and Strategies* (Washington, DC: Island Press, 1998).

**76.** Vlad Kacyznski, School of Marine Affairs, University of Washington, Seattle, presentation at Worldwatch Institute, 14 April 1994.

**77.** CFFA, op. cit. note 75; Buckley, op. cit. note 75; 1996 agreement from Milazzo, op. cit. note 8; for more details and history see Gareth Porter, "Euro-African Fishing Agreements: Subsidizing Overfishing in African Waters," in Scott Burns, ed., *Subsidies and Depletion of World Fisheries: Case Studies* (Washington, DC: WWF, April 1997).

**78.** Commoners from Bonnie J. McCay, "Sea Tenure and the Culture of the Commoners," in Cordell, op. cit. note 12; see also McGoodwin, op. cit. note 13.

**79.** Christopher L. Delgado and Anna A. McKenna, "Demand for Fish in Sub-Saharan Africa: The Past and the Future," *NAGA, The ICLARM Quarterly*, July–December 1997 (Supplement); Shakuntala Haraksingh Thilsted, Nanna Roos, and Nazmal Hassan, "The Role of Small Indigenous Fish Species in Food and Nutrition Security in Bangladesh," *NAGA, The ICLARM Quarterly*, July–December 1997 (Supplement).

**80.** Kent, op. cit. note 10; fish as share of animal protein and individual countries from FAO, *Marine Fisheries and the Law of the Sea*, op. cit. note 18, citing 1989 data that are still valid today; Meryl Williams, *The Transition in the Contribution of Living Aquatic Resources to Food Security, Food, Agriculture, and the Environment*, Discussion Paper 13 (Washington, DC: International Food Policy Research Institute, April 1996); 1 billion people from WRI, op. cit. note 10.

**81.** One third and 1992 per capita estimates based on "Fish and Fishery Products, Apparent Consumption," Appendix I, in FAO, *Yearbook of Fishery Statistics: Commodities*, vol. 81 (Rome: 1997). For purposes of consistency, unless otherwise indicated, all per capita data cited in this paper are for 1992, the most recent year available for apparent consumption statistics from FAO. Per capita supply available is calculated on the basis of 1992 population data given in Appendix I. For further discussion, please refer to FAO, *Food Balance Sheets, 1984–1986* (Rome: 1991).

**82.** FAO, op. cit. note 80.

**82.** Figure 4: regional per capita fish supplies from Appendix I, FAO, op. cit. note 81; share of animal protein from FAO, *Marine Fisheries and the Law of the Sea*, op. cit. note 18, citing 1989 data that are still valid today. Note: in 1992, people living in the former USSR and Eastern Europe had 10.3 kilograms available per person but data for fish as share of their total animal protein are unavailable for these countries, and therefore not included in the figure. Middle East includes Israel and Turkey; Latin America includes Caribbean, and Central and South America; Western Europe includes EU, plus Faeroe Islands, Iceland, Malta, Norway, and Switzerland.

**83.** Africa's 3 percent decline and Latin America's 8 percent decline in the 1980s from George Kent, "'Fish for the Poor': Competing with Chickens," *The Ecologist*, March/April, May/June, 1995; future from Williams, op. cit. note 80.

**84.** Growth in fish trade, Japanese imports, and high-priced species traded from Garcia and Newton, op. cit. note 7; Table 1 from *FAO Fishery Statistics Yearbook: Commodities*, vols. 45 and 81 (Rome: 1978 and 1997); Consumer Price Index (CPI) number used for 1970 is 0.247; for 1995 it is 0.971; from <http://www.orst.edu/ Dept/pol_sci/sahr/cpi96.htm>, viewed 2 February 1998; 32 percent from Williams, op. cit. note 80; John McQuaid,

"Fish Markets Going Global," *New Orleans Times-Picayune*, 30 March 1996.

**85.** Table 2 and 1,400 percent from FAO, op. cit. note 85, and CPI, op. cit. note 84; Thailand from Biksham Gujja and Andrea Finger-Stich, "What Price Prawn? Shrimp Aquaculture's Impact in Asia," *Environment*, September 1996; 1,000-fold is Worldwatch estimate based on FAO, *Aquaculture Production Statistics, 1986–1995*, op. cit. note 7.

**86.** Asia-Pacific Fishery Commission (APFC), *Regional Review of the Fisheries and Aquaculture Situation and Outlook in South and Southeast Asia* (Bangkok: FAO, 1996).

**87.** Kent, op. cit. note 10; earnings from FAO, op. cit. note 84, and CPI, op. cit. note 84.

**88.** India, Philippines, and quote from Kent, op. cit. note 10; global seafood prices from Mahfuzuddin Ahmed, "Fish for the Poor Under a Rising Global Demand and Changing Fishery Regime," *NAGA, The ICLARM Quarterly*, July–December 1997 (Supplement).

**89.** Kent, op. cit. note 83; malnutrition from Asian Institute of Technology, *Partners in Development: The Promotion of Sustainable Aquaculture* (Bangkok: January 1994).

**90.** Employer of last resort and 200 million from Williams, op. cit. note 80, and McGoodwin, op. cit. note 13; employer of only resort from R. W. Crowley and H. Palsson, "Rights Based Fisheries Management in Canada," *Marine Resource Economics*, summer 1992; world job estimates from FAO, *Numbers of Fishers 1970–1995*, FAO Fisheries Circular No. 929 (Rome: FAO, 1997); South and Southeast Asia estimates from "Plenary Discussion of Topic 1: Employment, Labour Productivity and Income of Small-Scale Fisheries," IPFC, op. cit. note 22; M'Bour from CFFA, op. cit. note 75; half the food from Talau-McManus and Kesner, op. cit. note 22.

**91.** Weber, op. cit. note 19.

**92.** Conner Bailey, "Discussion Guide: Employment, Labour Productivity and Income of Small-Scale Fisheries of South and Southeast Asia," in IPFC, op. cit. note 22; specific groups of people from Daniel Pauly, "Small-Scale Fisheries in the Tropics: Marginality, Marginalization, and Some Implications for Fisheries Management," in Pikitch, Huppert, and Sissenwine, op. cit. note 7.

**93.** Alain Le Sann, ed., *A Livelihood from Fishing: Globalization and Sustainable Fisheries Policies* (London: Intermediate Technology Publications, 1998); Kjell Rokke from "Giant Moves," *Samudra*, July 1996; Pescanova from "In the Red," *Samudra*, April 1995; Yamaha Motor Co., from "Yamaquaculture," *Samudra*, March 1997.

**94.** Japan from Michael L. Weber, "Effects of Japanese Government Subsidies of Distant Water Tuna Fleets," in Burns, op. cit. note 77; UNCLOS from Lisa K. Bostwick, "Empowering South Pacific Fishmongers: A New Framework for Preferential Access Agreements in the South Pacific Tuna Industry," *Law and Policy in International Business*, vol. 26, 1995; EU from Porter, op. cit. note 77.

**95.** Ibid; PNG from Milazzo, op. cit. note 8, citing "P.N.G. Ends Licensing Program," *Tuna Newsletter*, August 1996.

**96.** $229 million and portion paid by vessel owners from Porter, op. cit. note 77.

**97.** Mauritania from Milazzo, op. cit. note 8; coastline of Mauritania and 2.5 percent population growth from Hinrichsen, op. cit. note 75.

**98.** Porter, op. cit. note 77; Okechukwu C. Iheduru, "The Political Economy of Euro-African Fishing Agreements," *The Journal of Developing Areas*, October 1995.

**99.** Porter, op. cit. note 77; Milazzo, op. cit. note 8.

**100.** Data from Bostwick, op. cit. note 94; 5 percent from ADB, *The Pacific's Tuna: The Challenge of Investing in Growth, Pacific Studies Series* (Manila: April 1997).

**101.** Daniel MacMeekin, South Pacific fisheries law expert, MacMeekin and Woodworth, Washington, DC, e-mail to author, 11 May 1998; see also SPFFA from Gracie Fong, "Governance and Stewardship of the Living Resources: The Work of the South Pacific Forum Fisheries Agency," and Florian Gubon, "Steps Taken By South Pacific Island States to Preserve and Protect Ocean Resources for Future Generations," in Van Dyke, Zaelke, and Hewison, op. cit. note 13.

**102.** Greater benefits from restructuring from Bostwick, op.cit. note 94; what Japan pays Russia from Milazzo, op. cit. note 8; Sokimona from Wally Hiambohn, "PNG May Haul in Major Tuna Deal," *Post-Courier (PNG)*, 17 February 1998.

**103.** 2010 from FAO, *The State of World Fisheries and Aquaculture* (Rome: 1995); nonfood uses and fishmeal from Grainger and Garcia, op. cit. note 16; APFC, op. cit. note 86.

**104.** Carole Sugarman, "Aquaculture and Drugs: Wondering How Much Is Too Much," *Washington Post*, 14 August 1991; World Bank, "From 'Hunting' to Farming Fish: Rapid Production Increases Are Possible," press release (Washington, DC: 14 May 1995); Menakem Ben-Yami, "Aquaculture: The Importance of Knowing Its Limitations," *Ceres*, July–August 1986; Michael L. Weber, "So You Say You Want a Blue Revolution?" *Amicus Journal*, fall

1996; Holmes, op. cit. note 11.

**105.** George L. Chan, "Aquaculture, Ecological Engineering: Lessons from China," *Ambio*, November 1993; Ronald D. Zweig, "Freshwater Aquaculture in China: Management for Survival," *Ambio*, vol. 14, no. 2; for history see Clinton E. Atkinson, "People's Republic of China," in E. Evan Brown, ed., *World Fish Farming: Cultivation and Economics* (Westport, CT: Avi Publishing Company, 1977).

**106.** Vietnam, Philippines, and Ghana from Clive Lightfoot et al., "Aquaculture and Sustainability through Integrated Resources Management," *Outlook on Agriculture*, no. 3, 1993; for general discussion see P. Edwards, D.C. Little, and A. Yakupitiyage, "A Comparison of Traditional and Modified Inland Artisanal Aquaculture Systems," *Aquaculture Research*, vol. 28, 1997; for further discussion on potential see Roger S. V. Pullin and Mark Prein, "Fishponds Facilitate Natural Resources Management on Small-scale Farms in Tropical Developing Countries," in J.J. Symoens and J.C. Micha, *The Management of Integrated Freshwater Agro-Piscicultural Ecosystems in Tropical Areas* (Brussels: Technical Center for Agricultural and Rural Cooperation, Royal Academy of Overseas Sciences, 1995).

**107.** FAO, op. cit. note 103; growth rates and values from FAO, *Aquaculture Production Statistics 1984–1993* and *Aquaculture Production Statistics 1986–1995*, op. cit. note 7; catfish greater than veal, mutton, and lamb production combined from U.S. Department of Agriculture, Economic Research Service, LDP-AQS-7, *Aquaculture Outlook*, 5 March 1998.

**108.** Salmon and 85 percent freshwater from FAO, *Aquaculture Production Statistics 1986–1995*, op. cit. note 7; shrimp and mollusks from FAO, op. cit. note 103.

**109.** Peter N. Spotts, "Help for Fishermen from the Lab," *Christian Science Monitor*, 17 November 1995; Norway from Jason Clay, Senior Researcher, WWF, Washington, DC, discussion with author, 20 November 1997; "aquatic chicken" from David Dixon, "Superfish for the 21st Century?" *Far Eastern Agriculture*, September/October 1994.

**110.** Weber, op. cit. note 104; Chen Chunmei, "Aquaculture Meets Demand for Seafood," *China Daily*, 26 December 1995; C.K. Tseng, "Mariculture on Coastal China and Biodiversity," in Melvin N.A. Peterson, ed., *Diversity of Oceanic Life: An Evaluative Review*, Center for Strategic and International Studies (CSIS) Significant Issues Series, vol. 14, no. 12 (Washington, DC: 1992).

**111.** Gujja and Finger-Stich, op. cit. note 85; $6.3 billion from FAO, *Aquaculture Production Statistics 1986–1995*.

**112.** Michael L. Weber, *Farming Salmon: A Briefing Book* (San Francisco, CA: The Consultative Group on Biological Diversity, October 1997); value of salmon today from Jon Christensen, "Cultivating the World's Demand for Seafood," *New York Times*, 1 March 1997; share of 1995 world output and value for salmon and salmon and shrimp combined from FAO, *Aquaculture Production Statistics 1986–1995*, op. cit. note 7.

**113.** Profits from John Hambrey, "The Mangrove Questions," *Asian Shrimp News* (Bangkok: Asian Shrimp Culture Council, Second Quarter 1996); Matang from Jason Clay, *Market Potential for Redressing the Environmental Impact of Wild Captured and Pond Produced Shrimp* draft (Washington, DC: WWF, 21 November 1996).

**114.** Conditions for shrimp and price of conversion from Clay, op. cit. note 113; Gulf of Fonseco from Sarah Gammage, *Estimating the Returns to Mangrove Conversion: Sustainable Management of Short Term Gain?* Discussion Paper 97-02 (London: International Institute for Environment and Development (IIED), August 1997); Howard LaFranchi, "Chilean Can't See the Native Forests for the Woodchips," *Christian Science Monitor*, 8 January 1997.

**115.** Water needs from Clay, op. cit. note 113; Malcolm C.M. Beveridge, Lindsay G. Ross, and Liam A. Kelly, "Aquaculture and Biodiversity," *Ambio*, December 1994; water needs and problems of waste and water see Rebecca Goldburg and Tracy Triplett, *Murky Waters: Environmental Effects of Aquaculture in the US* (New York: Environmental Defense Fund, 1997).

**116.** Clay, op. cit. note 113; data from coastal villages in India and Bangladesh from Solon Barraclough and Andrea Finger-Stich, *Some Ecological and Social Implications of Commercial Shrimp Farming in Asia,* World Wildlife Fund (Geneva: United Nations Research Institute for Social Development (UNRISD) and WWF, 1996).

**117.** Clay, op. cit. note 113; Weber, op. cit. note 104; Cambodia, Thailand, and South Vietnam from Peter Edwards, "Sustainable Food Production through Aquaculture," *Aquaculture Asia*, January–March 1997; growth in fishmeal demand from FAO, *State of World Fisheries and Aquaculture*, op. cit. note 7; Ratafia, op. cit. note 18.

**118.** Up to one ton from Beveridge, Ross, and Kelly, op. cit. note 115; Goldburg and Triplett, op. cit. note 115; J.L. Maclean, "Developing Country Aquaculture and Harmful Algal Blooms," in R.S.V. Pullin, H. Rosenthal, and J.L. Maclean, eds., *Environment and Aquaculture in Developing Countries* (Manila: ICLARM and Deutsche Gesellschaft für Technische Zusammenarbeit, 1993); BC example from David Ellis, *Net Loss: The Salmon Netcage Industry in British Columbia* (Vancouver: David Suzuki Foundation, 1996).

**119.** Escapes from Beveridge, Ross, Kelly, op. cit. note 115, and Ellis, op. cit. note 118; Norway from Walter Gibbs, "Fish-Farm Escapees Threaten Wild Salmon," *New York Times*, 1 October 1996; Puget Sound from John

Fleischman, "Muddying the Waters," *Audubon*, March–April 1997.

**120.** Chua Thia-Eng, "Environmental Management of Coastal Aquaculture Development," and M.J. Phillips, C. Kwei Lin, and M.C.M. Beveridge, "Shrimp Culture and the Environment: Lessons from the World's Most Rapidly Expanding Warmwater Aquaculture Sector," in Pullin, Rosenthal, and Maclean, op. cit. note 118; Gulf of Thailand and lifespan from Mike Hagler, *Shrimp, the Devastating Delicacy: The Explosion of Shrimp Farming and the Negative Impacts on People and the Environment* (Washington, DC: Greenpeace, February 1997); rate of land abandonment from Gujja and Finger-Stich, op. cit. note 85.

**121.** Honduras from "Choluteca, Honduras: A Step Further," *Mangrove Action Project Quarterly News*, winter 1996, and Hagler, op. cit. note 120; India from Vijay Menon and Jacob George, "Farms in Deep Water," *India Today*, 31 January 1997, and Tara Patel, "Judge Pulls Plug on Polluting Shrimp Farms," *New Scientist*, 21–28 December 1996.

**122.** FAO, Aquaculture Development, *FAO Technical Guidelines for Responsible Fisheries No. 5* (Rome: 1997); FAO, "The Kyoto Declaration and Plan of Action on the Sustainable Contribution of Fisheries to Food Security," background paper, 9 December 1995.

**123.** Ibid.

**124.** For further discussion on raising noncarnivorous fish see Goldburg and Triplett, op. cit. note 115.

**125.** Joint Group of Experts on the Scientific Aspects of Marine Environmental Protection, *Monitoring the Ecological Effects of Coastal Aquaculture Wastes* (Rome: 1996); Jason W. Clay, "Toward Sustainable Shrimp Aquaculture," *World Aquaculture*, September 1997.

**126.** FAO, op. cit. note 103.

**127.** SPFFA from Fong and Gubon, op. cit. note 101; IPHC from Elliott A. Norse, ed., *Global Marine Biological Diversity: A Strategy for Building Conservation into Decision Making*, Center for Marine Conservation, IUCN, WWF, UNEP, and World Bank (Washington, DC: Island Press, 1993), and from Bruce R. Turris, "Canada's Pacific Halibut Fishery: A Case Study of an Individual Quota Fishery," in Gimbel, op. cit. note 51.

**128.** ICCAT from Sylvia A. Earle, *Sea Change: A Message of the Oceans* (New York: G.P. Putnam Sons, 1995), and from Carl Safina, op. cit. note 3; "management nightmare," from Mace, op. cit. note 31; Norse, op. cit. note 127.

**129.** Hinrichsen, op. cit. note 75.

**130.** Elisabeth Mann Borgese, "The Process of Creating an International

Ocean Regime to Protect the Ocean's Resources," in Van Dyke, Zaelke, and Hewison, op. cit. note 13; Clif Curtis, Policy Advisor, Greenpeace, Washington, DC, "Abstract of Environmental/Conservation Community Statement in Support of U.S. Accession to the Law of the Sea Convention," 8 June 1995.

**131.** Unqualified obligations from Lee A. Kimball, "The Law of the Sea Convention and Marine Environmental Protection," *The Georgetown International Environmental Law Review*, vol. 7, no. 3, 1995; Hinrichsen, op. cit. note 75; United Nations, *The Law of the Sea: Official Text of the U.N. Convention on the Law of the Sea with Annexes and Index* (New York: U.N. Publication, 1983).

**132.** James Carr and Matthew Gianni, "High Seas Fisheries, Large-Scale Drift Nets, and the Law of the Sea," in Van Dyke, Zaelke, and Hewison, op. cit. note 13; U.N. General Assembly, "Environment and Sustainable Development: Large-scale Pelagic Drift-net Fishing and Its Impacts on the Living Marine Resources of the World's Oceans and Seas," Forty-Ninth Session, Agenda Item 89, A/49/469, 5 October 1994; pockets of resistance and use of driftnets from David J. Doulman, "An Overview of World Fisheries: Challenges and Prospects for Achieving Sustainable Resource Use," presentation at the Law of the Sea Institute, 30th Annual Conference, Al-Ain, United Arab Emirates, 20 May 1996.

**133.** FAO, op. cit. note 7; U.N. Conference on Environment and Development, Agenda 21, Chapter 17, "Protection of the Oceans, All Kinds of Seas, Including Semi-Enclosed Seas, and Coastal Areas and the Protection, Rational Use and Development of their Living Resources," final, advanced copy, as adopted by the Plenary in Rio de Janeiro, 14 June 1992; U.N. Department for Policy Coordination and Sustainable Development, "Programme for the Further Implementation of Agenda 21: Adopted by the Special Session of the General Assembly, 23–27 June 1997," advanced unedited text, 1 July 1997.

**134.** FAO, *Code of Conduct for Responsible Fisheries* (Rome: 1995); Hinrichsen, op. cit. note 75; Deborah Hargreaves, "Environmental Groups Attack Voluntary Fishing Code," *Financial Times*, 17 March 1995.

**135.** U.N. Non-Governmental Liaison Service, "UN Conference on Straddling and Highly Migratory Fish Stocks: Final Negotiating Session," *Environment and Development File* (New York: August 1995), 16 million tons from Anne Platt, "High Seas Fishing," *World Watch*, September–October 1995; Table 3 based on Mace, op. cit. note 31.

**136.** States accountable for fleets from David R. Teece, "Global Overfishing and the Spanish-Canadian Turbot War: Can International Law Protect the High-Seas Environment?" *Colorado Journal of International Environmental Law and Policy*, no. 1, 1997; difficulty from Charles Higgenson, "The Law of the Sea Convention and the Protection of Fisheries," *The Georgetown*

*International Environmental Law Review*, no. 3, 1995; Satya N. Nandan, "UN Takes a Big Step to Conserve Fish Stocks," *Environmental Conservation*, autumn 1995; Moritaka Hayashi, "Enforcement by Non-Flag States on the High Seas Under the 1995 Agreement on Straddling and Highly Migratory Fish Stocks," *The Georgetown International Environmental Law Review*, no. 1, 1996; for more discussion on compliance provisions see Giselle Vigneron, "Compliance and International Environmental Agreements: A Case Study of the 1995 United Nations Straddling Fish Stocks Agreement," *The Georgetown International Environmental Law Review*, no. 2, 1998.

**137.** Michael Sutton, "Top Fishing Nations Drag Feet on UN Fish Stocks Agreement," press release (Washington, DC: WWF, 25 November 1997).

**138.** A. Charlotte de Fontaubert, David R. Downes, and Tundi Agardy, *Biodiversity in the Seas: Implementing the Convention on Biological Diversity in Marine and Coastal Habitats*, IUCN Environmental Policy and Law Paper No. 32, Marine Conservation and Development Report, published jointly by the Center for International Environmental Law, IUCN, and WWF, 1996; John Waugh, "The Global Policy Outlook for Marine Biodiversity Conservation," *Global Biodiversity*, vol. 6, no. 1; Suzanne Iudicello, "Protecting Global Marine Biodiversity," in William J. Snape III, ed., *Biodiversity and the Law* (Washington, DC: Island Press, 1996).

**139.** Ibid.

**140.** No mention of subsidies from Gareth Porter, "Natural Resource Subsidies and International Policy: A Role for APEC," *Journal of Environment & Development*, September 1997; few specifics and Diouf quote from Hargreaves, op. cit. note 134.

**141.** Iudicello, op. cit. note 138; Doulman, op. cit. note 132; Ellen Hey, "The Precautionary Approach: Implications of the Revisions of the Oslo and Paris Conventions," *Marine Policy*, July 1991; Ellen Hey, "The Precautionary Principle," *Marine Pollution Bulletin*, January 1993.

**142.** Mace, op. cit. note 31.

**143.** Know how to operate a boat from "News Round-Up: Canada," *Samudra*, July 1997; Pinkerton and Weinstein, op. cit. note 14; Svein Jentoft and Bonnie McCay, "User Participation in Fisheries Management: Lessons Drawn from International Experiences," *Marine Policy*, no. 3, 1995.

**144.** Pinkerton and Weinstein, op. cit. note 14; McGoodwin, op. cit. note 13; Mace, op. cit. note 31.

**145.** FAO, *Integration of Fisheries into Coastal Area Management*, FAO Technical Guidelines for Responsible Fisheries No. 3 (Rome: 1996); Callum M. Roberts, "Ecological Advice for the Global Fisheries Crisis," *Trends in Ecology and Evolution*, January 1997.

**146.** Robert S. Pomeroy et al., *Impact Evaluation of Community-Based Coastal Resource Management Projects in the Philippines,* (Manila: Fisheries Co-management Project at ICLARM and North Sea Center, June 1996); for further discussion of Philippines see Department of Environment and Natural Resources et al., *Legal and Jurisdictional Guidebook for Coastal Resource Management in the Philippines,* CRMP Document No. 8-CRM/1997 (Manila: CRMP, 1997).

**147.** Peter J. Auster and Nancy L. Shackell, "Fishery Reserves," in J.G. Boreman et al., eds., *Northwest Atlantic Groundfish: Perspectives on a Fishery Collapse* (Bethesda, MD: AFS, in press); Table 4: Apo Island and Saba Islands from de Fontaubert, Downes, and Agardy, op. cit. note 140; Apo Island also from Linda Bolido and Alan White, "Reclaiming the Island Reefs," *People & the Planet,* no. 2, 1997; Sri Lanka from Piyasena Ganewatte, "Collaborative Management at Rekawa Lagoon, Sri Lanka," *Intercoast Network: Mangrove Edition,* Special Edition no. 1, March 1997, and from Lynne Zeitlin Hale, "Putting an End to Coral Mining," *People & the Planet,* no. 2 , 1997; Belize from Sue Wells, "Protecting a Marine Wonderland," *People & the Planet,* no. 2, 1997; Texas from Alverson et al., op. cit. note 40; 20 percent from Karen F. Schmidt, "'No-Take' Zones Spark Fisheries Debate," *Science,* 25 July 1997, and Jane Lubchenco, "The Blue Report: Launching Marine Conservation Biology," presentation at the 1997 Annual Meeting of the Society for Conservation Biology, University of Victoria, British Columbia, 6 June 1997; 10 million tons from FAO, *State of World Fisheries and Aquaculture,* op. cit. note 7.

**148.** Gimbel, op. cit. note 51; differences from Neal D. Black, "Balancing the Advantages of Individual Transferable Quotas Against Their Redistributive Effects: The Case of *Alliance Against IFQs v. Brown,*" *Georgetown International Environmental Law Review,* vol. 9, 1997; Michael De Alessi, *Fishing for Solutions,* Institute for Economic Affairs, Studies on the Environment No. 11 (London: 1998).

**149.** Birgir Runolfsson, "Fencing the Oceans: A Rights-Based Approach to Privatizing Fisheries," *Regulation,* summer 1997; Resources for the Future, "Rights-Based Fishing: Transition to a New Industry," *Resources,* summer 1996.

**150.** Runolfsson, op. cit. note 149; Joshua John, *Managing Redundancy in Overexploited Fisheries,* World Bank Discussion Paper 240, Fisheries Series (Washington, DC: World Bank, 1994); Jed Greer, *The Big Business Takeover of US Fisheries: Privatizing the Oceans Through Individual Transferable Quotas (ITQs)* (Washington, DC: Greenpeace, April 1995); Leith Duncan, "Closed Competition: Fish Quotas in New Zealand," *The Ecologist,* March/April, May/June 1995; Ransom E. Davis, "Individually Transferable Quotas and the Magnuson Act: Creating Economic Efficiency in Our Nation's Fisheries," *Dickinson Journal of Environmental Law and Policy,* summer 1996. See also McLeod, op. cit. note 56.

**151.** Runolfsson, op. cit. note 149; Gimbel, op. cit. note 51; and Black, op. cit. note 148.

**152.** Black, op. cit. note 148; Newton and Fitzpatrick, op. cit. note 8; overview of ITQ programs in Organisation for Economic Co-operation and Development (OECD), *Toward Sustainable Fisheries: Economic Aspects of the Management of Living Marine Resources* (Paris: 1997).

**153.** Australia and global estimate from Milazzo, op. cit. note 8.

**154.** Christopher D. Stone, "The Crisis in Global Fisheries: Can Trade Laws Provide a Cure?" *Environmental Conservation*, 24, no. 2, 1997; Porter, op. cit. note 140; Milazzo, op. cit. note 8.

**155.** John Gates, Dan Holland, and Eyjolfur Gudmundsson, "Theory and Practice of Fishing Vessel Buyback Programs," in Burns, op. cit. note 77; Canada from "Trimming the Fleet: Can Wild Fisheries Recover?" op. cit. note 11; Europe from Caroline Southey, "Brussels Retreats on Cuts to Fishing Fleets," *Financial Times*, 11 October 1996; and Southey, "Move for Large Cuts in Fish Catches Undermined," *Financial Times*, 11 November 1996; for current situation in the United States see NMFS, "Establishing the Fishing Capacity Reduction Program," *SFA Update* (Sustainable Fisheries Act) August 1997.

**156.** Pomeroy et al., op. cit. note 146.

**157.** Indonesia from Robert S. Pomeroy, "Community-Based and Co-management Institutes for Sustainable Coastal Fisheries Management in Southeast Asia," *Ocean and Coastal Management*, no. 3, 1995; 30,000 islands and comparison to size of continental United States from Jeffrey Mervis, "Indonesia Opens the Door for Global Climate Studies," *Science*, 5 December 1997; 7.5 million fishers in India from Anne E. Platt, "It's About More Than Sea Cucumbers," *World Watch*, May/June 1995, and John Kurien, "Impact of Joint Ventures on Fish Economy," *Economic and Political Weekly*, 11 February 1995; Chile from Friedland, op. cit. note 6.

**158.** Namibia e-mail message from Les Clark, advisor, Ministry of Fisheries and Marine Resources, Namibia, on Fishfolk, electronic conference, 14 March 1998; Norway from Mace, op. cit. note 31.

**159.** U.S. from NMFS, *Magnuson-Stevens Fishery Conservation and Management Act, as Amended through 11 October 1996*, NOAA Technical Memorandum NMFS-F/SPO-23 (Silver Spring, MD: December 1996); "NOAA Using Congressional Tools to Combat Overfishing and Bycatch," NOAA News Release, 16 January 1998; turtle devices from Alverson et al., op. cit. note 40; Gulf of Mexico from Ted Williams, "The Turtle Gulf War," *Audubon*, September–October 1995; 60 percent from FAO, *State of World Fisheries and Aquaculture*, op. cit. note 7.

**160.** Barber and Pratt, op. cit. note 39; Chris Carr, Western Pacific Regional Fishery Management Council, Honolulu, Hawaii, Report from the South Pacific Forum Fisheries Agency, Sixth Technical Consultation on Vessel Monitoring Systems, 13–15 November 1996, Nadi, Fiji.

**161.** Precautionary management from Ludwig, Hilborn, and Walters, op. cit. note 56, Rosenberg et al., op. cit. note 56, and FAO, *Precautionary Approach to Capture Fisheries and Species Introductions*, FAO *Technical Guidelines for Responsible Fisheries No. 2* (Rome: 1996); ministries from Pauly, op. cit. note 92.

**162.** J.W. McManus, et al., *ReefBase Aquanaut Survey Manual* (Manila: ICLARM, 1997).

**163.** Models from P. Lehodey, et al., "El Niño Southern Oscillation and Tuna in the Western Pacific," *Nature*, 16 October 1997; Botsford, Castilla, and Peterson, op. cit. note 44; Steve Nadis, "Fishing for Data," *Technology Review*, January/February 1998.

**164.** Michael Sutton, "The Marine Stewardship Council: New Hope for Marine Fisheries," *NAGA, The ICLARM Quarterly*, July 1996; Ehsan Masood, "Fish Industry Backs Seal of Approval," *Nature*, 29 February 1996.

**165.** Barber and Pratt, op. cit. note 39; for the economic and ecological effects of shark fin trade, see Debra A. Rose, "An Overview of World Trade in Sharks and Other Cartilaginous Fishes," *Species in Danger Series*, TRAFFIC Network and Species Survival Commission (Cambridge, U.K.: WWF-UK and IUCN, 1996).

**166.** Public involvement from Michael Sutton, "Reversing the Crisis in Marine Fisheries: the Role of Non-Governmental Organizations," presentation at the Law of the Sea Institute, 30th Annual Conference, Al-Ain, United Arab Emirates, 20 May 1996.

# Worldwatch Papers

No. of Copies

***Worldwatch Papers by Anne Platt McGinn***

_____142. **Rocking the Boat: COnserving Fisheries and Protecting Jobs** by Anne Platt McGinn
_____129. **Infecting Ourselves: How Environmental and Social Disruptions Trigger Disease** by Anne E. Platt

---

_____141. **Losing Strands in the Web of Life: Vertebrate Declines and the Conservation of Biological Diversity** by John Tuxill
_____140. **Taking a Stand: Cultivating a New Relationship with the World's Forests** by Janet N. Abramovitz
_____139. **Investing in the Future: Harnessing Private Capital Flows for Environmentally Sustainable Development** by Hilary F. French
_____138. **Rising Sun, Gathering Winds: Policies to Stabilize the Climate and Strengthen Economies** by Christopher Flavin and Seth Dunn
_____137. **Small Arms, Big Impact: The Next Challenge of Disarmament** by Michael Renner
_____136. **The Agricultural Link: How Environmental Deterioration Could Disrupt Economic Progress** by Lester R. Brown
_____135. **Recycling Organic Waste: From Urban Pollutant to Farm Resource** by Gary Gardner
_____134. **Getting the Signals Right: Tax Reform to Protect the Environment and the Economy** by David Malin Roodman
_____133. **Paying the Piper: Subsidies, Politics, and the Environment** by David Malin Roodman
_____132. **Dividing the Waters: Food Security, Ecosystem Health, and the New Politics of Scarcity** by Sandra Postel
_____131. **Shrinking Fields: Cropland Loss in a World of Eight Billion** by Gary Gardner
_____130. **Climate of Hope: New Strategies for Stabilizing the World's Atmosphere** by Christopher Flavin and Odil Tunali
_____128. **Imperiled Waters, Impoverished Future: The Decline of Freshwater Ecosystems** by Janet N. Abramovitz
_____127. **Eco-Justice: Linking Human Rights and the Environment** by Aaron Sachs
_____126. **Partnership for the Planet: An Environmental Agenda for the United Nations** by Hilary F. French
_____125. **The Hour of Departure: Forces That Create Refugees and Migrants** by Hal Kane
_____124. **A Building Revolution: How Ecology and Health Concerns Are Transforming Construction** by David Malin Roodman and Nicholas Lenssen
_____123. **High Priorities: Conserving Mountain Ecosystems and Cultures** by Derek Denniston
_____122. **Budgeting for Disarmament: The Costs of War and Peace** by Michael Renner
_____121. **The Next Efficiency Revolution: Creating a Sustainable Materials Economy** by John E. Young and Aaron Sachs
_____120. **Net Loss: Fish, Jobs, and the Marine Environment** by Peter Weber
_____119. **Powering the Future: Blueprint for a Sustainable Electricity Industry** by Christopher Flavin and Nicholas Lenssen
_____118. **Back on Track: The Global Rail Revival** by Marcia D. Lowe
_____117. **Saving the Forests: What Will It Take?** by Alan Thein Durning
_____116. **Abandoned Seas: Reversing the Decline of the Oceans** by Peter Weber
_____115. **Global Network: Computers in a Sustainable Society** by John E. Young
_____114. **Critical Juncture: The Future of Peacekeeping** by Michael Renner
_____113. **Costly Tradeoffs: Reconciling Trade and the Environment** by Hilary F. Fronoh
_____110. **Gender Bias: Roadblock to Sustainable Development** by Jodi L. Jacobson
_____106. **Nuclear Waste: The Problem That Won't Go Away** by Nicholas Lenssen
_____105. **Shaping Cities: The Environmental and Human Dimensions** by Marcia D. Lowe

**_____Total copies (transfer number to order form on next page)**

## PUBLICATION ORDER FORM

_____ ***State of the World:* $13.95**
The annual book used by journalists, activists, scholars, and policymakers worldwide to get a clear picture of the environmental problems we face.

_____ **Worldwatch Library: $30.00 (international subscribers $45)**
Receive *State of the World* and all six Worldwatch Papers as they are released during the calendar year.

_____ ***Vital Signs:* $12.00**
The book of trends that are shaping our future in easy to read graph and table format, with a brief commentary on each trend.

_____ **WORLD WATCH magazine subscription: $20.00 (international airmail $35.00)**
Stay abreast of global environmental trends and issues with our award-winning, eminently readable bimonthly magazine.

_____ **Worldwatch Database Disk Subscription: $89.00**
Contains global agricultural, energy, economic, environmental, social, and military indicators from all current Worldwatch publications including this Paper. Includes a mid-year update, and *Vital Signs* and *State of the World* as they are published. Can be used with Lotus 1-2-3, Quattro Pro, Excel, SuperCalc and many other spreadsheets.
**Check one:** ____ **IBM-compatible** ____ **Macintosh**

_____ **Worldwatch Papers—See list on previous page**
**Single copy: $5.00** • 2–5: $4.00 ea. • 6–20: $3.00 ea. • 21 or more: $2.00 ea. (Call Vice President for Communications at (202) 452-1999 for discounts on larger orders.)

$4.00* Shipping and Handling *($8.00 outside North America)*

**minimum charge for S&H; call (800) 555-2028 for bulk order S&H*

_____ **TOTAL (U.S. dollars only)**

**Make check payable to Worldwatch Institute**

1776 Massachusetts Ave., NW, Washington, DC 20036-1904 USA

Enclosed is my check or purchase order for U.S. $________

☐ AMEX ☐ VISA ☐ MasterCard ______________________________
Card Number Expiration Date

______________________________
**signature**

______________________________
**name** **daytime phone #**

______________________________
**address**

______________________________
**city** **state** **zip/country**

**phone: (202) 452-1999 fax: (202) 296-7365 e-mail: wwpub@worldwatch.org**
**website: www.worldwatch.org**

Wish to make a tax-deductible contribution? Contact Worldwatch to find out how your donation can help advance our work.